Springer

Berlin
Heidelberg
New York
Barcelona
Budapest
Hongkong
London
Mailand
Paris
Santa Clara
Singapur
Tokio

Chemodynamics and Environmental Modeling

S. Trapp, M. Matthies

Chemodynamics and Environmental Modeling

An Introduction

Part 1: Textbook
Stefan Trapp
Michael Matthies

Part 2: CemoS User's Manual
Guido Baumgarten
Bernhard Reiter
Sven Scheil
Stefan Schwartz
Jan-Oliver Wagner

 Springer

Authors:
Dr. Stefan Trapp
Prof. Dr. Michael Matthies
University of Osnabrück
Institute of Environmental Systems Research
Artilleriestraße 34
D-49069 Osnabrück / Germany

CemoS was written by:
Guido Baumgarten, Bernhard Reiter, Sven Scheil,
Stefan Schwartz and Jan-Oliver Wagner
Institute of Environmental Systems Research
University of Osnabrück
Artilleriestr. 34
D-49069 Osnabrück / Germany
E-Mail CemoS@usf.Uni-Osnabrueck.de

Includes Chemical Exposure Model System CemoS on Disk

ISBN-13: 978-3-642-80431-1   e-ISBN-13: 978-3-642-80429-8
DOI: 10.1007/978-3-642-80429-8

Die Deutsche Bibliothek – CIP-Einheitsaufnahme
Chemodynamics and environmental modeling: an introduction;
with tables / textbook: S. Trapp ; M. Matthies. CemoS user's manual: G. Baumgarten ; B. Rei-
ter ; S. Scheil ; S. Schwartz ; J.-O. Wagner. - Berlin ; Heidelberg ; New York ; Barcelona ;
Budapest ; Hongkong ; London ; Mailand ; Paris ; Santa Clara ; Singapur ; Tokio : Springer
Dt. Ausg. u. d. T.: Dynamik von Schadstoffen – Umweltmodellierung mit CemoS
ISBN 3-540-63096-1
Buch 1998, Diskette. Chemical exposure model system CemoS. 1998

Typesetting with LaTeX : PTP - Protago • TeX • Production, Berlin
Cover: Struve & Partner, Heidelberg
SPIN: 10574988   52/3020 - 5 4 3 2 1 0 - Printed on acid-free paper

Preface

The estimation of chemical exposure, 'exposure modeling', is becoming increasingly important. There is no common concept or nomenclature. Comprehensive books about the underlying mathematical and physico-chemical principles are scarce. This book is dedicated to all those interested in chemical exposure prediction, but have not yet found an introduction. This includes system scientists, ecologists, chemical engineers, biologists, mathematicians, earth scientists and others, generally environmentally-oriented scientists. Knowledge of several disciplines is required, but there is no need to worry: this book is an introduction to the subject. All models are explained in detail, step by step, with one common concept. The simpler rather than the more complex structured models have been selected. To help with the understanding of the calculations, examples of applications are given. Exercises of different level are added to each Chapter, with solutions given at the end of the book. A list of literature for further reading is given at the end.

The book is based on lessons in 'Applied Systems Sciences' given from November 1992 until March 1997 at the University of Osnabrück. The manifold contributions of the students are hereby acknowledged. The 'CemoS' program (Chemical Exposure Model System) was developed between 1993 and 1996. This project was carried out by the following students: Guido Baumgarten, Bernhard Reiter, Sven Scheil, Stefan Schwartz and Jan-Oliver Wagner. The first edition of this book was in German (Trapp and Matthies 1996). This English version has been updated and some errors were eliminated, the CemoS version 1.10 with English user interface and updated chemical database is included and a new chapter on 'good modeling practice' has been added.

Osnabrück, August 1997   Stefan Trapp and Michael Matthies

Contents

Units and Abbreviations

<u>Masses</u>

kg   : kilogram
mg   : milligram
µg   : microgram $= 10^{-6}$ gram
ng   : nanogram $= 10^{-9}$ gram
pg   : picogram $= 10^{-12}$ gram
fg   : femtogram $= 10^{-15}$ gram
mol : moles

<u>Times</u>

s   : second
d   : day
a   : year

<u>Concentrations</u>

usually mass/volume, e.g. kg/m^3

ppm : parts per million $= mg/kg$
ppb : parts per billion $= µg/kg$
ppt : parts per trillion $= ng/kg$

<u>Others</u>

J   : Joule $= Nm$, unit of energy
Pa : Pascal $= N/m^2$, unit of pressure and fugacity
K   : Kelvin $= °C -273.15$, absolute temperature

Indices

A	:	atmosphere, air
B	:	bulk soil
d	:	distribution (K_d-value)
d	:	dry
E	:	effluent
f	:	fraction
F	:	fish
g	:	gaseous
i, j, l, n	:	running indices, number
l	:	liquid
L	:	leaves
M	:	matrix, e.g. soil or sediment matrix
O	:	n-octanol
P	:	plant
p	:	particle bound
Pa	:	particle (suspense in water, aerosols in air)
R	:	roots
S	:	sediment
t	:	total
w	:	wet
V	:	volatility
W	:	water
Xy	:	xylem

Indices and Symbols

A	:	area (m^2)
A	:	System matrix
a_{ij}	:	element of matrix, line i and row j
b	:	correction exponent for differences between lipids and octanol; 1 (fish), 0.95 (leaves), 0.77 (roots)
B	:	width of a river (m)
BCF	:	bioconcentration factor, concentration in biota to concentration in water
C	:	concentration, SI-unit kg/m^3; others, e.g. ppm, mg/kg
C(x,y,t)	:	concentration at the coordinates x and y at time t
CR	:	Courant-number $= u\,\Delta t/\Delta x$; determines the stability of the finite difference approximation
d	:	deposition ($kg\,m^{-2}\,s^{-1}$ or $kg\,m^{-2}\,a^{-1}$), Chapter 8
D	:	diffusion or dispersion coefficient ($m^2\,s^{-1}$)
D	:	dose kg/a, Chapter 8
Df	:	dilution factor, Chapter 6
D_{ij}	:	transfer coefficient in the fugacity approach ($mol\,h^{-1}\,Pa^{-1}$), Chapter 5
D_L	:	longitudinal dispersion coefficient in flow direction ($m^2\,s^{-1}$)
EL	:	evaporation limit in soil (m^3 water / m^3 soil), Chapter 7
f	:	fugacity (Pa), Chapter 5
f	:	fraction
$f(x)$	:	function of x
F	:	correction factor for high wind speeds, Chapter 6
FC	:	field capacity of soil (m3 water / m3 soil), Chapter 7

g	:	conductance (m/s), reciprocal to resistance r
h	:	depth of a river (m)
H	:	Henry's law constant with dimension ($Pa \ m^3 \ mol^{-1}$)
H	:	effective release height, Chapter 8
i	:	inhalation rate (m^3/h or m^3/a)
I	:	input of compound (kg/s)
J	:	flux per unit area ($kg \ m^{-2} \ s^{-1}$)
k	:	rate constant, e.g. for exchange (s^{-1})
k_l	:	conductance of the liquid film (m/s or m/h)
k_g	:	conductance of the gas film (m/s or m/h)
K	:	partition coefficient between phases (kg/m^3 to kg/m^3)
K_{AW}	:	partition coefficient between amosphere and water, non-dimensional Henry's law constant
K_{OC}	:	partition coefficient between organic carbon and water
K_{gp}	:	partition coefficient between gas and particles
K_{OW}	:	partition coefficient between octanol and water
K_d	:	partition coefficient between soil matrix and water (g/g or L/kg)
K_V	:	volatilization rate (s^{-1} or h^{-1})
L	:	lipid content (mass lipid per mass biota)
L	:	length (m)
$\log$	:	decadic logarithm, $\log_{10}$
m	:	mass (kg)
M	:	molar mass (g/mole)
n	:	substance mass in moles
n	:	number
n	:	Freundlich exponent, Chapter 6
N	:	netto substance flux (kg/s)
OC	:	organic carbon content (g OC / g dry bulk soil or sediment)
$[OH]\cdot$	:	concentration of oxygen radicals in air (molecules/cm^3)
p	:	vapor pressure (Pa)
p_s	:	saturation vapor pressure (Pa)
P	:	precipitation (mm/d or m/s); particle content of water, Chapter 6 (g/m^3);

q	:	darcy velocity (m/d or mm/d)
Q	:	volume flux, e.g. of water (m^3/s)
r	:	transfer resistance (s/m), reciprocal to conductance g
R	:	universal gas constant $= 8.314 \, J \, mol^{-1} \, K^{-1}$
S	:	solubility in water (mol/m^3 or kg/m^3)
S	:	sedimentation (mm/a) in Eq. 6.16
$S(i)$	:	soluble amount of substance in soil layer i (kg/m^2), Chapter 7
$S_{out,j}$	:	amount of substance leached out of soil column in time period j (kg/m^2)
t	:	time (s)
$t_{1/2}$	:	half-life $(s) = \ln 2/\lambda$
T	:	temperature (K or °C)
T	:	tortuosity factor: Chapter 7
TSCF	:	transpiration stream concentration factor
u	:	flow velocity in x-direction (wind or water) (m/s)
u_*	:	shear velocity (m/s)
u	:	eigenvector of a matrix
v	:	velocity, e.g. v_{dep}: deposition velocity in Chapter 6: wind speed at 10 cm height
V	:	volume (m^3)
W	:	water content (g/g)
$W(i)$	:	amount of water in soil layer i (m^3 water / m^2 soil), Chapter 7
$W_{out,j}$	:	amount of water lost by leaching or evaporation in the soil column in time period j (m^3 water / m^2 soil), Chapter 7
W_p	:	wash out ratio of particles
x	:	coordinate, distance (m); in Eq. 4.5: amount of substance (mol)
X	:	concentration per m^3 of phase, Chapter 7
y	:	coordinate, transversal (m)
z	:	coordinate, vertical (m)
Z	:	fugacity capacity (mol m^{-3} Pa^{-1})

ε : total porosity (vol/vol)

Φ : fraction of neutral (nondissociated) species of a chemical

λ : degradation rate, usually first order (s^{-1}); also: eigenvalue of a matrix

λ'_{OH} : degradation rate of second order for reaction with OH-radicals in air ($cm^3 \, mol^{-1} \, s^{-1}$)

θ : water pore fraction (vol/vol)

ρ : density (kg/m^3, also g/cm^3)

σ_y, σ_z : standard deviation of the Gaussian distribution function in y- and z-direction (m)

τ_r : average residence time (s) = reciprocal of the elimination rate

Chapter 1:
Why Model Chemical Exposure?

Facts:

More than 100 000 chemicals are available on the EU market.
More than 4 000 chemicals with annual production greater than 10 tonnes.
More than 1 000 chemicals with annual production greater than 1 000 tonnes.

3 336 'new' chemicals in the European Union (1983–1996)
129   chemicals classified as dangerous for water (FRG 1992)
???   chemicals dangerous for ???

1.1
Introduction

The world production of synthetic organic chemicals is about 250 million metric tons per year (Korte 1992) or even much higher (Colborn et al. 1996). Chemicals are either destroyed by use and thus producing degradation compounds, or released into the sewers and the air, or dumped, or incinerated after use. A small, but increasing fraction is used again (recycled). Compounds released from the technosphere usually enter the biosphere.

Since the beginning of environmental discussion, the problems of chemicals in the environment resulting from anthropogenic action have played a major role. The influence on natural cycles (for example, depletion of the stratospheric ozone), as well as the emission of synthetic xenobiotic substances (e.g. PCB) or of byproducts (e.g. PAH and chlorinated dioxins) changes our environment. Effects of synthetic chemicals on humans and wildlife cover a broad range from acute mortality to cancer and possibly endocrine disruption (see, e.g., Colborn et al. 1996). As well as the emission of millions of tonnes of high volume products per year, those of highly toxic micropollutants are also of great importance.

More than 100 000 different chemical compounds are marketed in the EU, about 10 000 of which are of great economic significance. Approximately 2 000

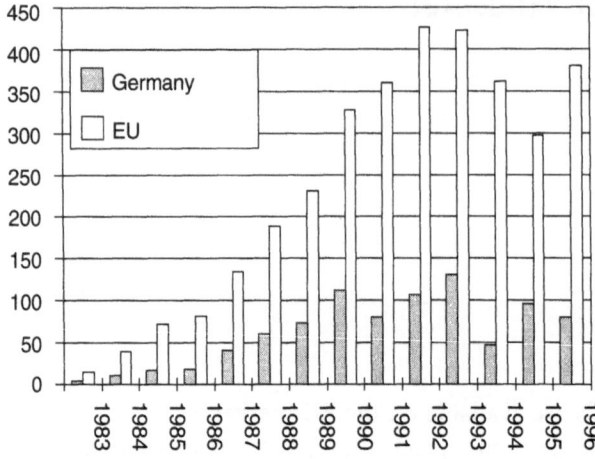

Fig. 1.1. Number of new chemicals on the German and European market

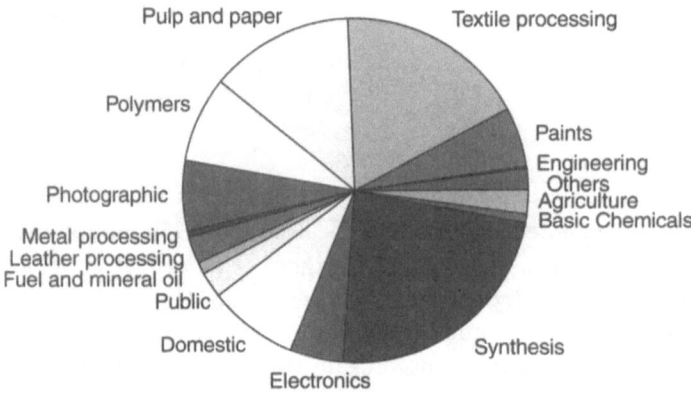

Fig. 1.2. Use categories of new chemicals in the EU (1983–1996)

High Production Volume Chemicals (HPV) are produced in volumes of more than 1 000 tonnes per year in the EU (Diderich and Ahlers, 1996). The transport and transformation processes are only known for a small fraction of these (GDCh 1988). 2 000 to 3 000 substances were identified alone in the atmosphere, these include many metabolites of the originally produced chemicals (Graedel et al. 1986). Additionally, every year, hundreds of new chemicals are brought to the market, alone in the EU, 3 336 new chemicals were notified between 1983 and 1996. They are used in very different categories, most of them for synthesis (23%), textile processing (18%) and pulp and paper (14%) (Figs. 1.1 and 1.2).

Scientific roots of environmental models

The need for tools to predict the environmental behavior of chemicals led to the development of mathematical models, which are designed to describe the transport and fate of substances under special environmental conditions. Exposure modeling of chemicals in environmental systems uses methods of systems analysis, model development and simulation of dynamic systems. Two different methods are identified:

- the *mechanistic* or deductive method, based on physical, chemical and biological processes and theories,
- the *empiric* or data based method which uses measured or observed parameters or time series.

Within the mechanistic process based model development, different approaches exist that have been developed by different scientific disciplines:

- the *hydrodynamic* or *flow mechanistic* approach focuses on ad/convection and dispersion, possibly in more than one spatial dimension, leading to partial differential equations that are usually numerically solved. Typical examples of this type are plume models for atmospheric transport (Chapter 8) or surface water simulation models (Chapter 6).
- the *reaction kinetic* approach which focuses on phase transfers or biochemical transformations of a substance. Spatial dependencies are often handled by building compartments. Ordinary differential equations then describe the substance behavior. Typical examples are the multimedia models (Chapter 5).

The first approach whose roots are in physics has been successfully used in meteorology and hydrology for many years. The reaction kinetic approach is derived from chemical engineering. Recently its use has increased in environmental modeling. Most exposure models incorporate both approaches to a varying degree.

The empiric method is very useful for those systems and processes that are too complex or too little understood for a detailed physico-chemical process description. Empirical relations found by experimental means may be used. A typical example is the sorption of organic chemicals to humic substances described by the K_{OC}.

During the last few years it has been increasingly recognized that for successful environmental protection the environment needs to be considered as a system. The influence of chemicals on environmental systems can only be explained by the interaction of many single processes. In particular, consideration of the dynamics of chemicals in the environment is required. Once released, chemicals are not longer controllable and may persist for a very long time before damage is apparent.

Spatial and time scales may differ greatly in environmental media. The large number of potentially environmentally hazardous compounds with varying properties and effects is another problem. This is particularly true for the class of organic chemicals. Much less is known about organics than about inorganics. Often, even the basic physico-chemical properties are unknown. Some organic chemicals have been better investigated and may serve as reference substances for environmental behavior to conclude the possible fate of others.

Risk assessment

When compounds are released into the environment they should not reach environmentally damaging concentrations. The concentration at the target is called exposure, a term derived from toxicology. The environmental hazard is the product of exposure and its effect. The probability of a hazard multiplied by the damage is the risk.

At present, substances are assessed by the ratio between "Predicted Environmental Concentration PEC" and "Predicted No-Effect Concentration PNEC". The PEC can be calculated from model simulations and may be used as assessment criteria:

$$\frac{PEC}{PNEC} > 1 ?$$

When the ratio is below one, there is no need for further testing or risk reduction measures. Otherwise, the question is whether further testing can lower the PEC/PNEC ratio. If the ratio is still above one, risk reduction measures are necessary (EU 1996, see also van Leeuwen and Hermens 1995).

Problematic seem mixtures (as they always occur in the environment): Almost nothing is known about the combined effect of chemicals, which may well be above the no-effect level. This is not considered in the current risk assessment methodology.

Application of exposure models

Exposure models have recently been developed for a number of purposes: the control of emissions into air (TA-Luft 1986); the estimation of the consequences of accidents and catastrophes; as part of the risk assessment of new and existing chemicals (EU 1996); the hazard assessment of contaminated sites (US-EPA 1996); the optimization of experiments and monitoring programs; and, finally, for life cycle assessment. Besides these diagnostic and protective motivations, scientific ones should also be mentioned. Models may help us to interpret and understand the complex interaction of processes. Inverse modeling can be used for the determination of parameters that cannot be measured directly, e.g. dispersion coefficients or environmental elimination half-life.

Unfortunately, the power of exposure models is often overestimated. Models certainly can help to give a range of results under plausible assumptions and thus may support decisions, but the environment is a complex and open system with extreme spatial and dynamic variability. Models of the environment are therefore limited. Simplification, neglections and idealizations are necessary, depending on the question, the problem, the data availability and quality.

All models described below are intended to assist the understanding of the principal behavior of chemicals in the environment. They are all deterministic models based on physical, chemical and biological theories, although they include empirical relations. The spatial and dynamic variability is always extremely simplified.

1.2
Good Modeling Practice

The estimation of environmental exposure has the following purposes (Wagner and Matthies 1996):

– to determine concentrations now and in the future in abiotic and biotic environmental segments,
– to understand fate processes and
– to recognize endangered ecosystems, populations or organisms.

The following steps are proposed to achieve these aims:

1. What are the purpose and scope?

Initially, the question to be answered needs to be clarified. For example, is there a short emission with high concentrations (accident) or a continuous long-term contamination? The environmental segment in question needs to be defined. The level of detail needs to be ascertained. Which time period is simulated? Which chemical or chemical class is considered?

2. Identification of underlying physical, chemical and biological processes

Processes to be considered are: release, transport such as ad/convection and dispersion or diffusion, chemical reaction, ad/desorption and transformation, biodegradation (and the formation of metabolites), and bioaccumulation.

3. Formulation of mathematical equations for single processes and model development

Processes regarded as important need to be translated into mathematical terms. Systems analysis and systems theory may help. The model will usually consist of a set of differential equations, where concentrations are the differential vectors, and processes describe their changes. An advanced simulation language may assist programming.

4. Data acquisition

Exposure models require substance data such as water solubility, and environmentally relevant data such as flow velocity. Some model parameters, such as sorption or degradation, depend on the environment and the chemical involved. In many cases, data is not directly available, e.g. bioaccumulation, but need to be estimated from other data by empirical relations.

5. Verification and calibration of the model structure

Verification of the model structure means testing of the underlying process description. This is done by using experimental data from selected reference chemicals, e.g. from microcosm. Calibration is fitting certain model parameters using these measurements.

6. Sensitivity analysis

Model results usually depend on a few input parameters. The sensitivity of parameters depends strongly on environmental and chemical data and must be determined for the specific problem. One way to do this is the variation of all input parameters (e.g. $\pm 10\%$) and the correlation with the output. Sensitive parameters show a high correlation.

7. Validation

The comparison between independently measured concentrations and the results of model simulations is usually called validation. Only a successful validation study leads to confidence in the predictive power of a model. However, the validity of the model strongly depends on its purpose (Rykiel 1996).

8. Uncertainty analysis

The uncertainty of model results also comes from the variance of input data (stochastic error), as well as from inadequate model structures, e.g. from oversimplification (systematic error). While stochastic errors can be estimated or calculated by stochastic methods (Monte-Carlo), this is more difficult for systematic errors. When more processes are considered to avoid oversimplification, the stochastical error increases as more input data are required. However, in many cases, missing processes cannot be described adequately or are unknown.

9. Decision and documentation

What was the purpose? What was the question to be answered? Model simulations may provide an answer to it, although in many cases, no clear answer is possible, either because of a lack of data or due to uncertainties. A report should always be prepared stating the question to be answered and the work that has been done. Without documentation and publication, all work must be repeated when a similar problem occurs.

Note: A model is a hypothesis that has to be tested by experiments. On the other hand, experiments are relatively useless if they do not lead to a hypothesis. With other words:

Model and experiment are like the egg and the hen!

Quality assurance

Due to the wide application and the possible economical and ecological impact of model calculations, the question of quality assurance arises. Within our program CemoS, we have implemented some 'quality assurance features' (see Part 2). We have found that at least the following things are necessary:

Transparency of the calculation; the user needs to know the purpose of the model. All equations used must be well described (and possibly cited). Together with CemoS, we deliver a text book and a manual. This manual is identically with the online help, that, however, is implemented as hyper text. Every point of the menu has its own link. Furthermore, a tutorial is given.

Complete documentation of input, estimates, output; together with the simulation results, the complete data set is exported. It contains all input data (with comments), all model results and names and all estimation functions that were used. The output is via the MIF (Model Interchange Format) that can be used as input file by CemoS. A speciality is that all data may be combined with one's own comment. This helps to avoid wrong data. Whenever data are changed, the comment must be renewed, and dependent variables are automatically updated (e.g. estimated parameters)

Error and warning ranges; during data input there may always be some mistakes made. It is nearly impossible to avoid this. However, CemoS does not allow one to enter physically impossible data. Unusually high or low values are only taken when confirmed. The warnings are saved and given with the output.

SI-units throughout; many errors occur during data recalculation, e.g. from non-metric to the metric system. CemoS uses SI-units throughout. This is sometimes unusual, for example, when concentrations are given in kg per cubic meter. But in fact, this avoids errors that often occur.

Analytical solutions to avoid numerical errors were used whenever possible (in fact, all models except the BUCKETS model are based on analytical solutions of the differential equations).

Validation studies; as pointed out, the successful proof of the applicability of a model helps one to rely on it. The authors have carried out (and still doing so) several validation studies to some of the models. They are cited in the text, and some examples are shown.

Limitations of applicability; due to regressions used, limited process consideration, lack of data or simply lack of knowledge, every model has limitations concerning its applicability. It is necessary to remind the user of this limitations. CemoS gives warnings when data are out of the regression range of an equation.

(E-mail)consulting; additional to the online help, users may contact the developers via e-mail. When the program is started, the address is the first thing that appears.

Tutorial; a model cannot be better than it's user. This text book is intended as an introduction into environmental modeling. External courses may be organized.

Updates; even with very careful programming and checking, mistakes may occur. Data in the data base may become out of date; models might evolve to take new scientific knowledge into account. This is why users may order updates (via FTP or on disk). The current English version (August 1997) is identical to the German version 1.10.

The user may notice some more details on quality assurance (e.g. accuracy) when using the program.

Closely related with the question of quality assurance is the standardization of models and parameter sets. Even small changes in the model (e.g. different estimation functions) and of the data sets (environmental and substance data) may significantly change the simulation results. Thus it is necessary to standardize the processes considered, and to compare the models using the same parameter set. Such a model comparison project was carried out by the *Society of Environmental Toxicology and Chemistry* SETAC (Cowan et al. 1996).

Open problems in exposure modeling

There are still many questions in exposure modeling which remain unanswered.

One of the more important ones is the question how results calculated for a homogeneous compartment may be interpreted for a strongly variable and structured environment. The problem is closely related to the question of 'scaling': how can results from microcosms and laboratory experiments be transferred to field situations.

There are several ways to proceed:

– a link between exposure models and geographic information systems. E.g., a number of small connected compartments may be used; currently the methodology is being developed to hold the necessary data within a geographical information system and transfer them automatically to the models (Matthies et al. 1997).

- stochastical variation of input parameters where the distribution of input parameters corresponds to the distribution of parameters in space and time.
- modeling the environment in its real detailed structure. This means dropping the assumption of homogeneous compartments and replacing ordinary by partial differential equations. For some environmental systems, the assumption of homogeneous compartments is generally not valid, e.g. for the unplugged soil. One of the current modeling activities is dealing with the behavior of compounds in the non-mixed multiphase system groundwater, soil, vegetation, air (Trapp and Matthies 1997).

Model simulations for prediction and interpretation of environmental contamination are a valuable supplement to other methods (laboratory studies, environmental monitoring) and have some economical and scientific advantages. They may help to avoid ecological damage. So it is most likely that there will be future developments in this area. This book may help to get an insight into the practice and methods of environmental modeling today. The software tool CemoS is primarily designed for didactical purposes, so you are invited to play with it, but it is also possible to use it as a professional tool. We hereby encourage the reader and user to contact us whenever problems occur and to send us publications about successful or unsuccessful applications.

Chapter 2:
Compartment Systems

This Chapter adresses the fundamental mathematical principles of the following models. It is not absolutely necessary for the use of the models but, nevertheless, it is the basis of model development.

2.1
Compartments

The environment is a spatially extremely structured dynamic system of abiotic and biotic components. Environmental factors interacting are e.g., wind, water, sun light, gravitation and temperature. Understanding and predicting the fate of chemicals in the environment requires a sketch of this complex system: a model.

An exposure model describes the main processes determining the environmental fate of a chemical and thus gives a summary of its environmental behavior. A frequently and successfully applied approach is the division of the environment into compartments. These compartments are assumed to be homogeneously mixed and to exchange chemical substances and energy between them (Jacquez, 1972). The detailed inner structure of a compartment is neglected. Compartments have a defined geometry, and therefore defined volumes, densities and masses. Compartments can also be called 'boxes', 'reservoirs', 'tanks' or 'pools', depending on the system under investigation.

Compartmental systems consist of a number of connected compartments (Fig. 2.1). The boxes represent compartments and the arrows represent exchanges or fluxes. Closed compartments only exchange between themselves, open compartments interact via input or output with their surroundings.

Compartment system analysis is closely related to the theory of dynamic systems (e.g. Luenberger 1979). The concepts of system theory can be applied, in particular to the mathematical methods.

Water, soil and air are the three main environmental compartments. The following assumptions are made for exposure modeling:

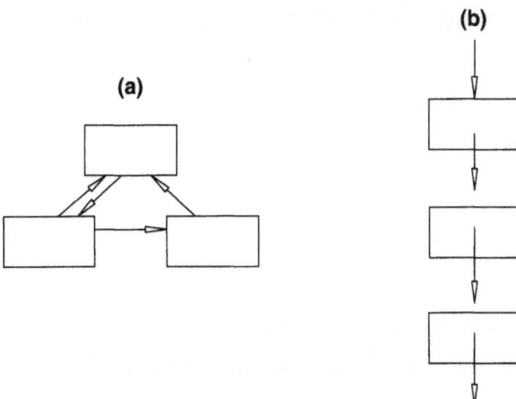

Fig. 2.1. Closed (a) and open (b) system with three compartments

1. Environmental compartments can be defined in such a manner as to represent phases or mixtures of phases in a thermodynamic sense.
2. Rules and laws of chemical equilibrium and kinetics can be applied to environmental systems.
3. Feedback of effects on organisms on the chemicals' fate is neglected.
4. No mixtures, only single compounds are considered. Interactions between components of mixtures are therefore also not considered.
5. Usually, only molecular-dispersively dissolved compounds and no separate phases of compounds are considered.

These assumptions are not trivial. Soil, e.g., is a mixture of the mineral phase, organic components, soil solution, gas phase and biota. However, a model structure can often be found that fulfills this purpose. The quality of a model is not measured by the number of model equations, but by the gain in knowledge (Veerkamp and Wolff 1996). This implies that the number of system components – the compartments – should be as small as possible. The model should only include the considerably important processes. It should also require a minimum of data and be comparable with experimental results.

2.2
Mass Balance / Linear Differential Equations

The law of mass conservation is the basis for modeling substance fluxes (1st law of thermodynamics). For compartments this means: the change of mass, $\Delta m(t)$, in the time interval Δt within the compartment is equal to source minus sink[1]:

$$\Delta m(t) = \{\text{source}(t) - \text{sink}(t)\}\Delta t \qquad (2.1)$$

---

1 This is similar to the profit in an economic context: profit $= d\$/dt = \{$expenditure $-$ costs$\}$ per time period.

Source and sink are variable with time. The resulting new mass at time $t = t_0 + \Delta t$ with $m_0 = m(t_0)$ is

$$m(t) = m_0 + \Delta m(t) = m_0 + \{\text{source}(t) - \text{sink}(t)\}\Delta t \tag{2.2}$$

It follows for continuous processes

$$\lim_{\Delta t \to 0} \{\Delta m(t)/\Delta t\} = dm(t)/dt = \text{source} - \text{sink} \tag{2.3}$$

If the source is independent of m(t) and the sink is proportional to m(t) (donor controlled), then

$$\text{source} = I(t) = \text{time-dependent input (mass/time)}$$

$$\text{sink} = -k\, m(t), \quad \text{where} \quad k = \text{elimination rate (time}^{-1})$$

$$\Delta m(t)/\Delta t = (m(t) - m_0)/\Delta t$$

and thus

$$\lim_{\Delta t \to 0} (\Delta m(t)/\Delta t) = dm(t)/dt = I(t) - k\, m(t) \tag{2.4}$$

In the environment, concentrations rather than masses are measured.

With concentration = mass m per volume V it follows for constant volume

$$dC(t)/dt = I(t)/V - k\, C(t) \tag{2.5}$$

Equations 2.4 and 2.5 are linear differential equations of the first order with constant coefficient k and an external input function $I(t)$. The solution depends on the initial value $C(t_0) = C_0$ and the input function $I(t)$. The general solution of the initial value problem with the integration variable t' is

$$C(t) = C_0 \exp[-k(t - t_0)] + 1/V \int_{t_0}^{t} \exp[k(t - t')]\, I(t')dt' \tag{2.6}$$

If the integral in Eq. 2.6 cannot be solved analytically, a numerical integration method has to be used (see below). Without input and with $t_0 = 0$ a *homogeneous linear differential equation of the first order* arises with the solution

$$C(t) = C_0 \exp(-kt) \tag{2.7}$$

Radioactive decay and many chemical reactions follow this equation. Half-life $t_{1/2}$ is defined as

$$t_{1/2} = \ln 2/k = 0.693k^{-1} \tag{2.8}$$

Residence time τ_r is the reciprocal of the elimination rate k:

$$\tau_r = k^{-1} = t_{1/2}/0.693 \tag{2.9}$$

With constant input $I(t) = I_0$ the solution of the *inhomogeneous linear differential equation of the first order* is

$$C(t) = C_0 \exp(-kt) + I/(V\,k)[1 - \exp(-kt)] \tag{2.10}$$

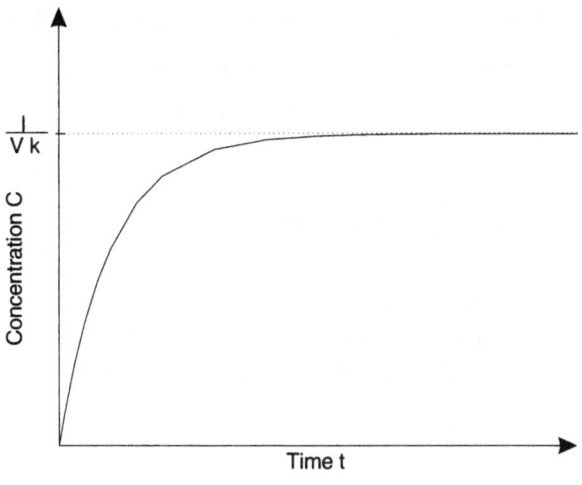

Fig. 2.2. Time course of concentration $C(t)$ in the mixing tank

Example 2.1: Mixing tank

A chemical flows with a constant input I_0 into a mixing tank of (constant) volume V. There, it is homogeneously mixed and leaves the tank at rate k.

The basic equation is (analogous to Eq. 2.5):

$$dC(t)/dt = -k\,C(t) + I_0/V \qquad (2.11)$$

The analytical solution with $C_0 = 0$ is

$$C(t) = I_0/(V\,k)[1 - \exp(-kt)] \qquad (2.12)$$

The steady-state concentration $C(t \to \infty)$ depends on the quotient between input and elimination rate at given volume, $C(t \to \infty) = I_0/(V\,k)$.

2.3
Differential Equation Systems

There now follow some basic mathematics about linear differential equation systems. The only model that uses this is that of chapter 10. If you have an aversion to 'math', you might consider going on to the next chapter.

Systems of compartments

In a system of compartments, the exchanges between the compartments play an additional role. All compartment systems leading to n linear differential equations (n compartments) with constant coefficients are now considered. **Vectors** are given in bold print.

$$dm_1(t)/dt \ = a_{11}m_1(t) + a_{12}m_2(t) + \ldots + a_{1n}m_n(t) + I_1(t) \qquad (2.13)$$
$$dm_2(t)/dt \ = a_{21}m_1(t) + a_{22}m_2(t) + \ldots + a_{2n}m_n(t) + I_2(t)$$
$$dm_n(t)/dt \ = a_{n1}m_1(t) + a_{n2}m_2(t) + \ldots + a_{nn}m_n(t) + I_n(t)$$

For compartment i:

$$dm_i(t)/dt = \sum_{j=1, j \neq i}^{n} a_{ij}m_j(t) + a_{ii}m_i(t) + I_i(t) \qquad (2.14)$$

The first term on the right is the exchange with all other compartments j. The second term includes all sinks from compartment i (e.g. degradation, a_{ii} is negative). The third term on the right represents all external sources into compartment i.

Written as a matrix:

$$\dot{\mathbf{m}}(t) = \mathbf{A}\mathbf{m}(t) + \mathbf{I}(t) \qquad (2.15)$$

$\dot{\mathbf{m}}(t) \quad = \quad$ vector of differential quotient $dm_i(t)/dt$ (change of mass with time)

$A \qquad = \quad$ compartmental or systems matrix (although bold, not a vector)

$\mathbf{m}(t) \quad = \quad$ vector of masses, state variable

$\mathbf{I}(t) \qquad = \quad$ input vector

Steady-state solutions

Steady-state means that the system does not change with time. The fluxes are in equilibrium, sources and sinks are balanced. The differential term of the equation system is zero:

$$\mathbf{A}\mathbf{m} + \mathbf{I} = 0 \qquad (2.16)$$

A linear equation system results which is easily solved by the usual methods, e.g. by Gaussian elimination. Linear differential equation systems come to a steady-state for $t \to \infty$.

Example 2.2: Aquarium

Let us consider the "aquarium", a two-compartment system, Fig. 2.3.

The fish takes up a substance from the surrounding water with the rate constant k_1 and secretes it with a clearance rate constant k_2. The mass balance is:

$$dm_W(t)/dt = -k_1 m_W(t) + k_2 m_F(t) \qquad (2.17a)$$
$$dm_F(t)/dt = k_1 m_W(t) - k_2 m_F(t) \qquad (2.17b)$$

with m_F and m_W = masses in fish and water, resp.

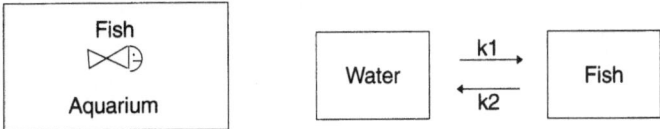

Fig. 2.3. Two-compartment "aquarium" system with fish and water

Assuming no substance is lost, we have a closed system. In a steady-state $(dm_F(t)/dt = dm_W(t)/dt = 0)$ the following concentration ratio between fish and water applies; * denotes steady-state conditions:

$$dm_W/dt \quad = \quad -k_1 m_W{}^* + k_2 m_F{}^* = 0 \qquad (2.18)$$
$$k_1 m_W{}^* = k_2 m_F{}^*$$
$$k_1/k_2 = m_F{}^*/m_W{}^* = V_F C_F{}^*/(V_W C_W{}^*)$$
$$k_1 V_W/(k_2 V_F) = k_1'/k_2' =$$
$$= \quad C_F{}^*/C_W{}^* = \text{bioconcentration factor BCF}$$

The concentration ratio at $t \to \infty$ is thus determined by the quotient between uptake and clearance rate constant at a given volume. This is frequently used for the experimental determination of the bioconcentration factor. The concentration in the water is measured at different times. From the concentration course (initially contaminated water and clean fish), the uptake rate k_1' is calculated. The fish is then brought into clean water. The clearance rate k_2' is determined by the increase in concentration there.

Solution of special compartment systems

In some special cases the differential equation system can be directly solved by repeated integration.

Example 2.3: Cascade

A liquid flows through a cascade of two tanks (Fig. 2.4).
    The mass balance is therefore

$$dm_1(t)/dt = -k_1 m_1(t) \qquad (2.19a)$$

$$dm_2(t)/dt = k_1 m_1(t) - k_2 m_2(t) \qquad (2.19b)$$

The initial substance masses are $m_1(0) = m_0$ and $m_2(0) = 0$; with $t_0 = 0$, $m_1(t)$ is:

$$m_1(t) = m_0 \exp(-k_1 t) \qquad (2.20)$$

For $m_2(t)$ we get:

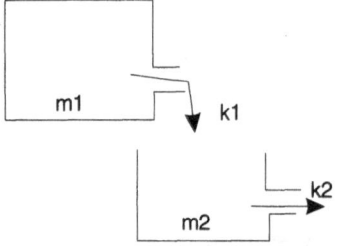

Fig. 2.4. Cascade

$$dm_2(t)/dt = k_1 m_0 \exp(-k_1 t) - k_2 m_2(t) \tag{2.21}$$

From this it follows by integration for $m_2(t)$:

$$m_2(t) = k_1 m_0/(k_2 - k_1)[\exp(-k_1 t) - \exp(-k_2 t)] \tag{2.22}$$

For a cascade of length three (= three compartments) we get:

$$
\begin{aligned}
m_3(t) \quad = \quad k_1 k_2 m_0 \Bigg\{ & \frac{\exp(-k_1 t)}{(k_2 - k_1)(k_3 - k_1)} \\
& + \frac{\exp(-k_2 t)}{(k_1 - k_2)(k_3 - k_2)} + \frac{\exp(-k_3 t)}{(k_1 - k_3)(k_2 - k_3)} \Bigg\}
\end{aligned}
\tag{2.23}
$$

and accordingly for n linearly connected compartments:

$$m_n(t) = \prod_{i=1}^{n-1} k_i m_0 \left\{ \sum_{i=1}^{n} \frac{\exp(-k_i t)}{\prod\limits_{j=1,j\neq i}^{n} (k_j - k_i)} \right\} \tag{2.24}$$

General solution method for linear differential equation systems

Let us consider a system of linear differential equations of the form

$$\dot{\mathbf{m}}(t) = \mathbf{A}\mathbf{m}(t) + \mathbf{I}(t)$$

$\dot{\mathbf{m}}(t)$ = vector of differential quotients dm_i/dt
$\mathbf{A}$ = compartmental or system matrix
$\mathbf{m}(t)$ = vector of masses
$\mathbf{I}(t)$ = input vector

The corresponding homogeneous differential equation system (homogeneous in this context means: $\mathbf{I} = 0$) only has a non-vanishing solution if:

$$\det(\mathbf{A} - \lambda\mathbf{E}) = 0 \quad \text{(E is the unit matrix)} \tag{2.25}$$

The eigenvalues λ of $\mathbf{A}$ are the roots of the 'characteristic polynomial'. The eigenvectors $\mathbf{u}$ of $\mathbf{A}$ are the non-vanishing solutions of

$$(A - \lambda E)\mathbf{u} = 0 \tag{2.26}$$

If the $n \times n$ matrix A has n independent eigenvectors of the eigenvalues $\lambda_1, \ldots, \lambda_n$ ($\lambda_1, \ldots, \lambda_n$ need not be different), then the general solution of the linear differential equation system is (Braun 1983, Jacquez 1972):

$$\mathbf{m}(t) = c_1 \exp(\lambda_1 t)\mathbf{u}_1 + c_2 \exp(\lambda_2 t)\mathbf{u}_2 + \ldots + c_n \exp(\lambda_n t)\mathbf{u}_n \tag{2.27}$$

The constants c_i are calculated from the initial conditions ($t = 0$). In the case of multiple eigenvalues, independent eigenvectors can also always be found, see Braun (1983).

Analogously to the one-dimensional case (Eq. 2.5), a solution can be found for the inhomogeneous equation system with $\mathbf{I}(t)$:

$$\mathbf{m}(t) = \exp(At)\mathbf{m}_0 + \int_0^t \exp[A(t - t')]\mathbf{I}(t')dt' \tag{2.28}$$

Instead of the scalar exponent, we have here the matrix exponential. If $\mathbf{I}(t) = 0$, the homogeneous solution is

$$\mathbf{m}(t) = \exp(At)\mathbf{m}_0 \tag{2.29}$$

The calculation of the matrix exponential $\exp(At)$ can be found in the relevant text books on systems science (Luenberger 1979) or on differential equations (Braun 1983). For other input vectors (pulses, periodic input), solutions can be taken from systems theory (Luenberger 1979).

2.4
Numerical Solution Methods for Ordinary Differential Equations

Advantages and disadvantages of numerical methods

Numerics is a special field within mathematics. It deals with the solution of equations, i.e. finding the most exact value. Numerical methods are iterative, i.e. they proceed step by step. This is extremely troublesome by 'hand', but numerical schemes are frequently applied on computers.

Some of the advantages and disadvantages of analytical solution methods are: Analytical solutions are easier to handle than numerical methods, they are stable and avoid numerical errors. The resulting computer codes can run on small computers or even on hand calculators. The great disadvantage is that analytical solutions can only be found for special conditions. Variable initial and boundary conditions can rarely be considered.

Numerical solution methods do not have these limitations. This allows advanced model formulations, but then data and computation required increase considerably. A significant problem is the accuracy of the numerical solution, which requires a careful selection and design of the appropriate numerical scheme.

Here, we will briefly consider the basics of numerical solution schemes. For a detailed description, please refer to the literature on numerics.

All linear differential equation systems with constant coefficients and constant input can be solved analytically. Most of the models presented in the latter are solved in this way. Our main purpose is the understanding of environmental systems behavior and not the simulation of real cases. Analytical solutions are more transparent for the reader and thus are didactically of greater importance.

One-step Euler scheme

The one-step *Euler* scheme is the easiest of all numerical schemes for the solution of ordinary differential equations (but by no means the most accurate). It is suitable for the communication of the principle.

Example: Reaction of the first order, no input (Eq. 2.5)

$$dC(t)/dt = -kC(t) \tag{2.30}$$

Substitute the differential quotient dC/dt with the difference quotient $\Delta C/\Delta t$. By definition, dC/dt is the limit of $\Delta C/\Delta t$ for $\Delta t \to 0$. We get:

$$\Delta C(t)/\Delta t = -kC(t) \tag{2.31}$$

and

$$\Delta C(t) = -kC(t)\Delta t \tag{2.32}$$

For $C(t + \Delta t)$:

$$C(t + \Delta t) = C(t) + \Delta C(t) = C(t) - kC(t)\Delta t \tag{2.33}$$

and again:

$$\begin{aligned} C(t + \Delta t + \Delta t) &= C(t + \Delta t) + \Delta C(t + \Delta t) \\ &= C(t + \Delta t) - kC(t + \Delta t)\Delta t \ \ldots \end{aligned}$$

The selection of the time step Δt is critical for this solution scheme. Δt must be small enough to avoid deviations of the linear approximation. But if the time step is too small, rounding errors occur. There is no exact criterion for the best time step with the Euler scheme. In any case, it should be smaller than the inverse of the fastest rate. Usually, when the time step is varied and the results do not change, the selected time step may be used. Problems occur mainly with rapid changes. In this case, a very small Δt needs to be selected.

Four-step Runge-Kutta scheme

Although the Euler scheme is easy, it is rarely used. A classical and frequently applied method is the four-step *Runge-Kutta* scheme.

At first, four approximations for $C(t+\Delta t)$ are calculated; one partial result is used for the next, and so on. The result for the total time step is then calculated from the four partial steps. In our example with the first order reaction kinetics, this is:

Runge-Kutta scheme 4th Order

first step

$$s1 = -kC \tag{2.34a}$$

second step

$$s2 = -k(C + 1/2\Delta t\, s1) \tag{2.34b}$$

third step

$$s3 = -k(C + 1/2\Delta t\, s2) \tag{2.34c}$$

fourth step

$$s4 = -k(C + 1/2\Delta t\, s3) \tag{2.34d}$$

and complete step

$$\Delta C = \Delta t(s1 + 2s2 + 2s3 + s4)/6 \tag{2.35}$$

and analogous to 2.33:

$$C(t + \Delta t) = C(t) + \Delta C$$

and so on for all following time steps.... . A criterion for the time step selection of the Runge-Kutta scheme is:

$$testrk = |(s2 - s3)/(s1 - s2)| < 0.05 \tag{2.36}$$

The general scheme of the method is:

Be $y = y(x)$ a solution of the initial value problem $dy = f(x, y)$, $y(x_0) = y_0$. At given positions $x_1, x_2, x_3 \ldots$, approximations $y_1, y_2, y_3 \ldots$ of the exact values $y(x_1), y(x_2), y(x_3)$ can be found. The difference $\Delta x_v = x_{v+1} - x_v$ ($v = 0, 1, 2, \ldots$) is the step size h. The four parts of the solution are $y_{v+1}^I, y_{v+1}^{II}, y_{v+1}^{III}$ and y_{v+1}^{IV}. Then

$$y_{v+1}^I \quad = y_v + hf(x_v, y_v) \tag{2.37a}$$

$$y_{v+1}^{II} \quad = y_v + hf(x_v + h/2, 1/2(y_v + y_{v+1}^I)) \tag{2.37b}$$

$$y_{v+1}^{III} = y_v + hf(x_v + h/2, 1/2(y_v + y_{v+1}^{II})) \tag{2.37c}$$

$$y_{v+1}^{IV} = y_v + hf(x_v + h, y_{v+1}^{III}) \tag{2.37d}$$

and

$$y_{v+1} = (y_{v+1}^I + 2y_{v+1}^{II} + 2y_{v+1}^{III} + y_{v+1}^{IV})/6 \tag{2.38}$$

For further details please refer to the literature on numerics of differential equations.

Comparison between Euler, Runge-Kutta scheme and analytical solution

The comparison between the solution methods is shown by the example of
Exercise 2.2 in Table 2.1. The selected time step is large. The Euler scheme
fails, but the Runge-Kutta scheme is only a few per cent away from the (exact) analytical solution. Both Euler and Runge-Kutta schemes can be used for
the integration of ordinary differential equations. The latter is advantageous,
although it involves slightly more effort for the programmer.

There may be cases when the Euler scheme may be superior to the Runge-Kutta scheme, in particular when coefficients of the matrix differ greatly, e.g.
between 100 and 10^{-5}. This case sometimes occurs when dynamics of chemicals in different environmental media are simulated (e.g. water and sediment).
More elaborate methods for the solution of such problems are described in
textbooks on numerical mathematics, e.g. Butcher (1987), Gear (1975), Lapidus
and Schiesser (1975) or Shampine and Gordon (1975).

Table 2.1. Degradation of trichloroethene (exercises on Chapter two); initial concentration
$C(0) = 100$mg/L, degradation rate constant $= 0.01d^{-1}$ and time step $\Delta t = 50$ days (very
long); testrk $= |(s2 - s3)/(s1 - s2)| = 0.25$ (should be < 0.05)

Time (days)	Euler	Runge-Kutta	Exact
50	50.0000	60.6771	60.6531
100	25.0000	36.8171	36.7879
150	12.5000	22.3395	22.3130
200	6.2500	13.5550	13.5335
250	3.1250	8.2247	8.2085
300	1.5625	4.9905	4.9787
350	0.7812	3.0281	3.0197

Exercises on Chapter Two

Exersise 2.1: Degradation Chain

Trichloroethene = Tri (C_2HCl_3) is a solvent used in metal working and textile
industry. Millions of tons of it have been produced world-wide, production
is decreasing. Today, it is found at hundred thousands of sites in groundwater, where it is a serious contaminant. Tri is biodegraded to (cis- and
trans-)dichloroethene at a rate constant of $\lambda = 0.01\ d^{-1}$, this at a rate constant of $0.11\ d^{-1}$ to vinyl chloride (carcinogenic!) and this at a rate constant of
$0.002\ d^{-1}$ to ethene (comparatively harmless). All rate constants are assumed
to be first order. They may vary considerably with environmental conditions.

$$\lambda = 0.01d^{-1} \qquad \lambda = 0.11d^{-1} \qquad \lambda = 0.002d^{-1}$$

Tri $\longrightarrow$ Di $\longrightarrow$ VC $\longrightarrow$ Ethene

a) If the initial concentration of Tri is 100 mg/L, what is the concentration after one year?
b) What is the concentration of vinyl chloride after one year?

(Note: neglect the fast step to dichlorethene).

Exersise 2.2: Ozone indoor

Indoor ozone (O_3) has a half-life of approximately 5 minutes. Let the volume of your living room be 50 m^3. You open the window. The air exchange is 2 m^3/min. On extreme days, the concentrations of ozone outside can be up to 300 $\mu g/m^3$ ozone.

a) What is the steady-state concentration in your living room? What is the concentration after opening the window for 5 min, 1 hour?
b) When is the steady-state reached?
c) When is the deviation from steady-state below 5%?
d) You sit in your living room and inhale 1 m^3/h air. Ozone reacts completely with your lungs. What dosage do you inhale in one hour?

Exersise 2.3: Aquarium

a) Write down the mass balance of a substance in water and fish for a closed aquarium!
b) What is the analytical solution for mass in water and fish, when an initial input into water occurs?

Chapter 3:
Transport and Transformation Processes

Up till now, we have considered interactions in a compartment system without looking at the underlying physico-chemical processes. In the environment, important processes are transport, e.g. flowing with a river and reaction, e.g. photodegradation. We now will consider these processes in more detail. At first, the physical transport processes of diffusion, ad/convection and dispersion are described, followed by chemical reaction processes.

3.1
Diffusion

Experiment: Put a droplet of ink into a glass of water. At first, the droplet is deep blue and small. But after a relatively short time, the water is homogeneously light blue. The reason is the undirected random movement of water and ink. Without any influence from outside, mixing occurs.

The microscopic movement of molecules by heat leads to mixing and to the disappearance of concentration gradients. Particles or molecules show a net flux from places with higher to places of lower concentration. This was first mathematically formulated by Fick (Fick's First Law). The process is called molecular diffusion (see, e.g., Atkins 1986).

In the one-dimensional case, the net substance flux N_{diff} (kg s^{-1}) through an area A (m^2) of thickness Δx (m) at a given concentration gradient ΔC(kg m^{-3}) is:

$$N_{diff} = J_{diff} A = -A D \Delta C / \Delta x \tag{3.1}$$

J_{diff} is the net substance flux through the unit area (kg s^{-1} m^{-2}), D is the diffusion coefficient (m^{2}s^{-1}) and $\Delta C/\Delta x$ is the concentration gradient (kg m^{-3}m^{-1}). The driving force for the diffusion are the concentration gradients. The minus sign is convention: flux is in a positive x-direction when the concentration gradient is negative.

Fick's First Law is only valid when

- the medium is isotrope (the structure of the medium and diffusion coefficient is the same in all directions)
- the flux by diffusion is perpendicular to the cross section area
- the concentration gradient is constant.

The diffusion coefficient D in gases is within the range of 10^{-5} to 10^{-4} m^2s^{-1}, in liquids about 10^{-9} m^2s^{-1} and in solids about 10^{-14} m^2s^{-1}. It can be seen that D primarily depends on the aggregate state, and diffusion in gases is most significant. The diffusion coefficient is proportional to temperature and inversely proportional to the molecule volume. This again is related to the root of the molar mass. The relation between the diffusion coefficients of two substances is therefore approximately (Tinsley 1979, Schwarzenbach et al. 1993):

$$D_i/D_j = \sqrt{M_j}/\sqrt{M_i} \tag{3.2}$$

M is the molar mass (g/mol), i and j are indices for two different chemicals. With Eq. 3.2, an unknown diffusion coefficient can be calculated approximately from known ones (compare Section 11.7, estimation of diffusion coefficients, Eq. 11.13).

The quotient between diffusion coefficient D and diffusion length Δx is conductance g (m/s), a measure of the exchange velocity. Conductance g is inverse to resistance r (s/m) (Gates 1980):

$$D/\Delta x = 1/r = g \tag{3.3}$$

If more than one resistance occurs, the total resistance is calculated from the Kirchhoff laws, similar to the electric resistance (Gates 1980): with resistances in series (one after another), the total resistance is the sum of the single resistances. When resistances are in parallel, the conductances are added.

Resistances in series:

$$r_{total} = r_1 + r_2 + \ldots + r_n \tag{3.4}$$

Resistances in parallel:

$$g_{total} = g_1 + g_2 + \ldots + g_n \tag{3.5}$$

and

$$r_{total} = 1/g_{total} \tag{3.6}$$

Fick's second law of diffusion

Diffusion not only occurs as an exchange between separated compartments, but also in spatially continuous phases, such as water and air (e.g. ink droplet in water). Now we consider the spatial dependencies explicitly, first of all for one dimension with x as a coordinate. The expression N_{diff} in Eq. 3.1 is equivalent to the change of mass in time interval Δt, i.e. the difference quotient $\Delta m/\Delta t$:

$$\Delta m/\Delta t = -A\,D\,\Delta C/\Delta x \tag{3.7}$$

With $C = m/V$, C now depends on x and t. This gives us a partial differential quotient, Fick's First Law in differential form:

$$\partial C/\partial t = -A/V\,D\,\partial C/\partial x \tag{3.8}$$

If now the concentration gradient $\Delta C/\Delta x$ itself changes with $\Delta x = -V/A$, the term $\Delta/\Delta x\,\Delta C/\Delta x$ results, and it follows

$$\partial C/\partial t = \Delta/\Delta x(D\,\partial C/\partial x) \tag{3.9}$$

Expressing this in differential form gives Fick's Second Law:

$$\partial C/\partial t = \partial/\partial x(D\,\partial C/\partial x) \tag{3.10}$$

The environment is an anisotropic medium, and D usually depends on x. For mathematical purposes, D is often assumed to be constant:

$$\partial C/\partial t = D\,\partial^2 C/\partial x^2 \tag{3.11}$$

For three dimensions x, y and z, the diffusion equation is (Crank 1970):

$$\frac{\partial C}{\partial t} = D_x \frac{\partial^2 C}{\partial x^2} + D_y \frac{\partial^2 C}{\partial y^2} + D_z \frac{\partial^2 C}{\partial z^2} \tag{3.12}$$

where D_x, D_v and D_z are the (constant) diffusion coefficients in x, y and z directions.

3.2
Dispersion

Molecular diffusion appears in all environmental media, e.g. in air, water and soil. Diffusion is a property of the molecule (movement by heat, 2nd law of thermodynamics). Other undirected mixing processes can be described with similar equations, although the reason is not molecular diffusion. Turbulence of the surrounding medium (random movement in one or all directions) causes a similar mixing process. It is called dispersion. Dispersion is a property of the surrounding medium and is independent of molecular properties. Dispersion is often orders of magnitudes faster than molecular diffusion. Formally, dispersion can be described like molecular diffusion in Eqs. 3.1 and 3.11:

$$N_{disp} = J_{disp}A = -D_{disp}\,A\,\Delta C/\Delta x \tag{3.13}$$
$$\partial C/\partial t = D_{disp}\,\partial^2 C/\partial x^2 \tag{3.14}$$

The molecular diffusion coefficient is replaced by the dispersion coefficient D_{disp}. During turbulent mixing, too, concentration gradients disappear. Dispersion is always connected to flow processes. Unlike diffusion, dispersion is unisotrop. The transversal dispersion coefficient (perpendicular to the flow direction) is usually much smaller than the longitudinal (in flow direction). During flow in porous media (e.g. soil or groundwater), the dispersion is caused by differences in flow length through pores of different size. In meteorology, the dispersion is called "eddy diffusion". The dispersion coefficient is an empirical parameter and is found by calibration, e.g. of tracer experiments.

3.3
Advection (Convection)

A substance contained in a flowing medium is co-transported. The process is called advection or convection. First of all we will again look at one dimension. A substance is dissolved in water with the concentration C. The water flows with current velocity u (m s^{-1}) through the cross-sectional area A (perpendicular to u):

$$N_{adv} = J_{adv} A = A u C \tag{3.15}$$

The concentration change in time by advective substance flux N_{adv} can be expressed as:

$$\partial C/\partial t = A/V u C \tag{3.16}$$

If C and u are not constant, but depend on x in the interval $\Delta x = -V/A$, then

$$\partial C/\partial t = -\Delta/\Delta x \, (uC) \tag{3.17}$$

Having a spatially continuous flow, the difference quotient can be replaced by the differential quotient

$$\partial C/\partial t = -\partial(uC)/\partial x \tag{3.18}$$

This is the continuity equation, based on the conservation of mass. In segments of the environment, the flow velocity may be assumed to be constant, then

$$\partial C/\partial t = -u \, \partial C/\partial x \tag{3.19}$$

Diffusion, dispersion and advection are universal processes. They appear in all environmental media. Their description is the basis of many transport models of chemicals in water, air and soil. The mass of substance decreases neither by diffusion, nor by dispersion or advection. They are pure transport processes.

3.4
Combination of Diffusion, Dispersion and Advection

The combination of diffusive/dispersive and advective transport terms results in the diffusion/dispersion-advection equation. With one dimension, where D is the sum of diffusion and dispersion coefficients:

$$\frac{\partial C}{\partial t} + u\frac{\partial C}{\partial x} = D\frac{\partial^2 C}{\partial x^2} \tag{3.20}$$

Basic solution of the diffusion-advection equation

The mathematical advantage of compartmentilization of the environment is the independence of spatial dimensions. Homogeneous mixing is assumed, leading to ordinary differential equations. But when both spatial and dynamic changes are considered, partial diffential equations result.

For the solution of partial differential equations, not only initial conditions (time dependency), but also boundary conditions (spatial dependency) need to be considered. Techniques for the analytical solution of partial differential equations are separation of variables or Laplace transformation (Crank 1970).

For the initial and boundary conditions

– the diffusion/dispersion coefficient is constant in space and time,
– the cross-sectional area A (m²) and the flow velocity u (m/s) are constant,
– input m_0 (kg) is at $x = 0$ and time $t = 0$,
– $C(\infty, t) = 0$ (no boundary)

the solution is

$$C(x, t) = \frac{m_0/A}{(4\pi Dt)^{1/2}} \exp\left[-\frac{(x - ut)^2}{4Dt}\right] \tag{3.21}$$

m_0 : substance input (kg) at $t = 0$,
A : cross-sectional area (m²),
t : flow time (s),
x : flow distance (m),
u : flow velocity (m s^{-1}),
D : sum of diffusion and dispersion coefficients (m² s^{-1}).

The advection is expressed as a transformation of the coordinate x with the flow distance $u \cdot t$.

The equation is very similar to the probability density of the normal distribution

$$f(x) = \frac{1}{\sigma(2\pi)^{1/2}} \exp\left[-\frac{(x - \mu)^2}{2\sigma^2}\right] \tag{3.22}$$

μ is the mean and σ the standard deviation of a normally distributed random variable.

Due to the central limit theorem, the normal distribution results from the addition of n ($n \rightarrow \infty$) independent stochastic events of the same order of magnitude (Sachs 1992). Diffusion is an independent stochastic movement of molecules, thus the resulting density (concentration) is normally distributed.

Example 3.1: Accident, dispersion

A truck falls into the River Rhine and spills its load of ten ton completely into the water; $m_0 = 10\,000\,\text{kg}$. The Rhine in this segment (near Cologne) has a width of approx. 333.3 m and a depth of 3 m, giving a cross-sectional area A of $1\,000\,\text{m}^2$. An approximate value of the dispersion coefficent is $500\,\text{m}^2/\text{s}$ (empirical value, $\gg$ molecular diffusion). First of all let us just consider dispersion (flow velocity $u = 0$). We calculate concentration C at x and t. The result is shown in Fig. 3.1.

The peak broadens due to dispersion. The mass is unchanged. The peak concentration decreases, but a wider segment of the river is affected. Of course, the example is completely fictitious, since dispersion does not exist without flow.

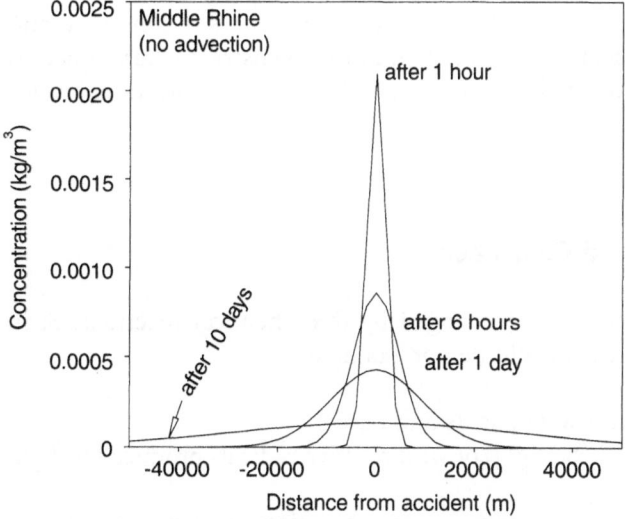

Fig. 3.1. Effect of dispersion

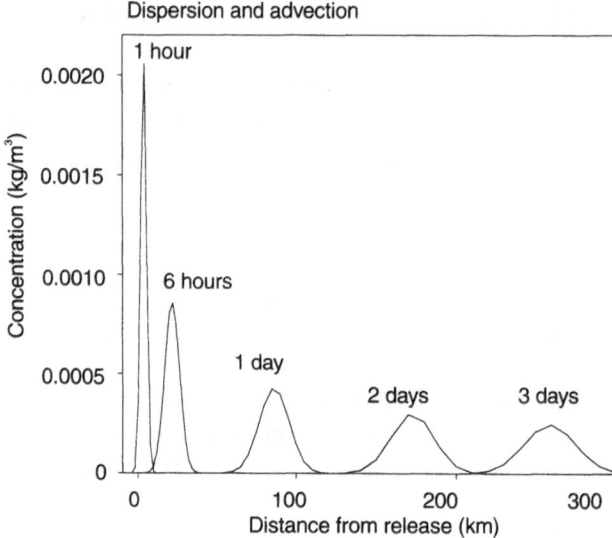

Fig. 3.2. Effect of dispersion plus advection

Example 3.2: Accident, dispersion plus advection

In rivers, dispersion, diffusion and advection occur simultaneously. At normal flow velocities, dispersion is by far larger than diffusion. Advection is most important. Now, flow velocity u is 1 m/s. The result is shown in Fig. 3.2. While the water flows downstream, the peak broadens. At the same time, the coordinate in flow direction changes due to advection. Again, the mass is conserved and the peak concentration decreases, but now it flows downstream. Subsequently, new river segments are affected downstream with lower concentration, but for a longer time period.

3.5
Reaction, Metabolism and Elimination

The mass of the chemical can be changed by (bio)chemical reactions. Such reactions in the environment could be, for example:

- hydrolysis (reaction with water, acids and bases)
- photolysis (reaction with sunlight, or with reactive radicals produced by light energy)
- biodegradation (reaction with enzymes or other biogenic compounds)
- oxidation (reaction with oxygen or burning)

In order to describe the time course of reactions, the velocity of a reaction is defined as concentration of a reaction partner transformed per time period. Let us consider a reaction of the type

Reaction partner k + reaction partner $l \rightarrow$ products

with concentrations C_k and C_l of the reaction partners. If the reaction velocity is independent of the concentrations of the reaction partners, the reaction is of zero order. If it depends on the concentration of only one partner C_k, the reaction is of the first order. If one partner is present in excess, the reaction type is usually of the first order. The reaction type is of the second order, when the reaction velocity is proportional to the product of the concentrations of both partners k and l. Reactions of the second order may occur when two partners have similar concentrations. Reactions of higher orders are very rare.

Equations

Zero order

$$dC/dt \;\; = \;\; -\lambda^0 \tag{3.23}$$
$$C(t) \;\; = \;\; C_0 - \lambda^0 t \tag{3.24}$$

λ^0: reaction rate constant zero order, unit is concentration per time.

first order

$$dC/dt \;\; = \;\; -\lambda C \tag{3.25}$$
$$C(t) \;\; = \;\; C_0 \exp(-\lambda t) \tag{3.26}$$

λ: here first order reaction rate constant with unit 1/time. When more reactions of first order type occur, the rates can be added:

$$\lambda_{\text{total}} = \lambda_1 + \lambda_2 + \lambda_3 + \ldots + \lambda_n \tag{3.27}$$

second order

$$dC_k/dt = -\lambda' C_k C_l \tag{3.28}$$

λ' is the second order rate constant with the unit 1/(time · concentration).

pseudo-first order

A reaction of the second order can be expressed as 'pseudo'-first order by multiplying the second order rate constant with the concentration of the reaction partner C_l:

$$\lambda = -\lambda' C_l \tag{3.29}$$

Michaelis-Menten kinetics

The velocity v of a enzymatically catalysed transformation (mass per time) depends on enzyme concentration, substrate concentration (C), the affinity of the enzyme to the substrate (K_m) and on the maximal velocity (v_{max}). At high substance concentrations, a saturation effect will occur. This describes the *Michaelis-Menten* equation:

$$v = (v_{max}C)/(K_m + C) \tag{3.30}$$

where K_m is the *Michaelis-Menten* constant (Schlegel 1981). It gives the substrate concentration at which the enzyme activity is half of the maximal velocity v_{max}. When $C \ll K_m$, the transformation velocity is approximately linear to C (first order reaction). When $C \gg K_m$, the transformation velocity is maximally v_{max} and independent of C (zero order reaction). Analogous are the *Monod kinetics* for microbial degradation.

3.6
Combination of Dispersion, Advection and Elimination

Let us consider the analytical solutions for advection plus dispersion and for first order degradation. The complete equation for all processes simultaneously results from the multiplication of both solutions:

$$c(x, t) = \frac{m/A}{\sqrt{(4\pi Dt)}} \exp\left[-\frac{(x - ut)^2}{4Dt}\right] \exp(-\lambda t) \tag{3.31}$$

$c(x, t)$: concentration of the substance at x and t (kg/m^3);
x : coordinate in flow direction (m);
t : time after release (s);
m : amount released (kg);
A : cross-sectional area (m^2);
D : dispersion coefficient (m^2/s);
u : flow velocity in x-direction (m/s);
λ : first order reaction rate constant (s^{-1})

Example 3.3: Accident with elimination

For the calculation of our truck accident ($m_0 = 10\,000$ kg, $A = 1000$ m^2, $D = 500$ m^{2}s^{-1}, $u = 1$ ms^{-1}) it is now assumed that the chemical released into the water is eliminated with a half-life of one day ($\lambda = 8 \cdot 10^{-6}$ s^{-1}), Fig. 3.3.

The peak again flows downstream and broadens. But now the mass decreases significantly. After 300 km (approx. 3.5 days), only 890 kg of the released 10 000 kg are left, since $m(t) = m_0 e^{-\lambda t}$.

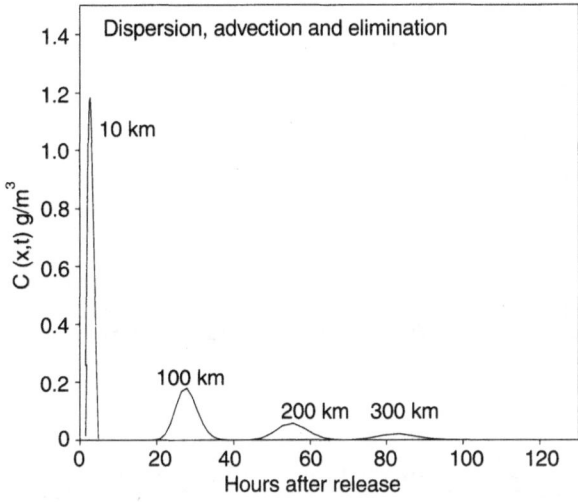

Fig. 3.3. Dispersion, advection and elimination

3.7
Numerical Solution of the Dispersion-Advection Equation

With the appropriate methods, partial differential equations can also be solved numerically. Among the several existing methods are the finite difference methods (FDM).

Finite difference method

The spatial coordinate (here x) is divided into segments. A constant concentration is assumed within these segments or at the nodes. The diffences between the nodes are calculated (finite difference method FDM) for discrete time steps Δt. A differential is replaced by a difference.

Example: a river is divided into segments. These segments may have different properties, due to the variability of the river. Segments are subdivided into boxes (Fig. 3.4).

We get intervals Δx and Δt (space and time). $C_{x,t}$ becomes $C_{i,j}$; i is the index for the spatial segments and j is the index for the time period. The differential quotients of the dispersion/diffusion-advection equation (Eq. 3.20) are replaced by the differences

$$\frac{\partial C}{\partial t} \approx \frac{C_{i,j+1} - C_{i,j}}{\Delta t} \tag{3.32}$$

$$u\frac{\partial C}{\partial x} \approx u\frac{C_{i,j} - C_{i-1,j}}{\Delta x} \tag{3.33}$$

$$D\frac{\partial^2 C}{\partial x^2} \approx D\frac{(C_{i+1,j} - C_{i,j}) - (C_{i,j} - C_{i-1,j})}{(\Delta x)^2} \tag{3.34}$$

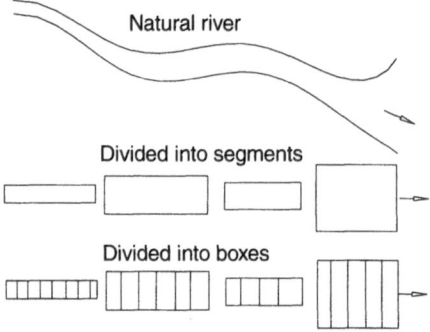

Natural river

Divided into segments

Divided into boxes

Fig. 3.4. From continuity to boxes

The complete equation is

$$\frac{\partial C}{\partial t} \approx \frac{C_{i,j+1} - C_{i,j}}{\Delta t} = D\frac{C_{i+1,j} - 2C_{i,j} + C_{i-1,j}}{(\Delta x)^2} - u\frac{C_{i,j} - C_{i-1,j}}{\Delta x} \quad (3.35)$$

The equation is solved for $C_{i,j+1}$. The concentration at time $j+1$ $(j = 1,\ldots,n)$ thus only depends on concentration at time j. The method is directed forwards, it is an explicit finite difference method. Concentrations $C(i,1)$, $i = 1,\ldots,m$ are the initial condition, and concentrations $C(n+1,j)$, $j = 1,\ldots,n$ are the lower boundary conditions.

The accuracy of the solution depends largely on the selection of Δx and Δt. The solution is only stable (not oscillating) when (Kinzelbach 1992):

1) Courant-criterion: Advective flow may not take within one time step Δt more out of a box as is in there:

$$CR = \Delta t\, u/\Delta x \leq 1 \quad (3.36)$$

   CR: Courant number
2) Neumann criterion: During one time step, concentration gradients between two boxes may not change direction:

$$D\,\Delta t/(\Delta x)^2 \leq 1/2 \quad (3.37)$$

3) Combination of both:

$$2\,\Delta t\, D/(\Delta x)^2 + \Delta t\, u/\Delta x \leq 1 \quad (3.38)$$

The maximal time step Δt results:

$$\Delta t \leq \Delta x/(2\,D/\Delta x + u) \quad (3.39)$$

One more criterion follows from the reaction term: $\Delta t \leq 1/\lambda$

Consideration of numerical dispersion

Stability is one precondition for the convergence of the scheme. It does not mean, however, that the solution is correct. One of the more obvious problems is the numerical dispersion. Just consider the advective part of the equation:

$$\partial C/\partial t + u\,\partial C/\partial x = 0$$

By difference approximation, we get

$$\Delta C/\Delta t + u\,\Delta C/\Delta x = 0$$

From $\Delta C/\Delta x = (C_i - C_{i-1})/\Delta x$ follows the concentration $C_{i,j+1} = C_{i,j} + \Delta C$ at $x = i$ and $t = j + 1$:

$$C_{i,j+1} = C_{i,j} - u/\Delta x\,(C_{i,j} - C_{i-1,j})\,\Delta t \tag{3.40}$$

In words: the concentration at $x = i$ at the next time period $j + 1$ is the concentration at this node at the momentary time period $C_{i,j}$, minus the substance flowing out from segment i into the downstream segment in this period, plus the substance flowing into it from the upstream segment $i - 1$.

A 'substance package' leaving the nodes i, j at time t will be at $t + \Delta t$ (with $\Delta t < \Delta xu$) between the nodes $(i, j + 1)$ and $(i + 1, j + 1)$. With the difference approximation, the package is divided into two segments, although it has not yet reached the second one. The process is similar to dispersion.

By use of a Taylor series development, Kinzelbach (1992, pp. 193–194) showed that the difference approximation does not solve the original equation, but the equation plus an additional artificial dispersion term:

$$\partial C/\partial t + u\partial C/\partial x = 1/2(u\Delta x - u^2\Delta t)\partial^2 C/\partial x^2$$

By selecting appropriate time and space steps, the artificial numerical dispersion can be made equal to the real physical dispersion:

$$D_n = 1/2(u\Delta x - u^2\Delta t) \tag{3.41}$$

For the numerical solution of the dispersion-advection equation with the FDM, the value of D_n needs to be subtracted from the input value of the longitudinal dispersion D_L.

For small values of Δx, the numerical dispersion vanishes, but at the same time, Courant number and Neumann criterion increases, and the solution becomes unstable and oscillates between $+\infty$ and $-\infty$.

Trick: In Eq. 3.40 we have not yet explicitly considered dispersion. We choose Δx so that D_n is equal to D_L:

$$\Delta x = 2D_L/u + u\Delta t \tag{3.42}$$

If CR is exactly 1, then the movement of the substance package through the time-space grid is described exactly by the i, j- nodes ($CR = 1$, $u\Delta t/\Delta x = 1$, $u\Delta t = \Delta x$). The solution is stable for $CR \leq 1$. If CR is equal to 1, we have the *exact* solution without (numerical) dispersion. However, we now need the numerical dispersion and choose $CR = u\Delta t/\Delta x = 3/4$, and then

$$\Delta x = 4/3\, u\Delta t \tag{3.43}$$

To get the value of the numerical dispersion D_n exactly equal to the real longitudinal dispersion D_L, Eq. 3.42 must hold. It follows for Δt:

$$\Delta t = 6D_L/u^2 \tag{3.44}$$

and Eq. 3.44 applied to Eq. 3.43 gives

$$\Delta x = 8D_L/u \tag{3.45}$$

Now the grid is fixed in space and time. For every node i, j, we can solve the dispersion/advection equation. We also can select a Δt or Δx, and calculate the other parameter, if necessary.

The resulting space-time grid is not suitable for all problems. The minimal space step Δx for small Δt is $2D_L/u$. For small-scale simulations (e.g. in soil), the grid is often not applicable. But it is a good method for flowing rivers.

Beware!

Besides oscillation and numerical dispersion, some further numerical problems may occur, e.g. 'overshooting' and rounding errors. Errors are often difficult to locate as long as the results look plausible. It is absolutely necessary to compare results with analytical solutions. This is the best way to identify and avoid mistakes.

Example 3.4: Alarm! On Sunday at 10 p.m. a tanker loaded with chemicals had an accident on the Rhine at km 430. One tonne of bromacil (a herbicide) was released. What is the concentration at the border to the Netherlands (km 830)?

Approximate average values of the Rhine: width 333.3 m, depth 3 m, flow velocity 1 m/s (volume flow 1 000 m³/s), $D_L = 500\,\mathrm{m^2 s^{-1}}$.

Select $CR = 0.75$:

$$\Delta t = 6D_L/u^2 = 6 \times 500\,\mathrm{m^2 s^{-1}}/(1\,\mathrm{ms^{-1}})^2 = 3\,000\,\mathrm{s}$$
$$\Delta x = 8D_L/u = 8 \times 500\,\mathrm{m^2 s^{-1}}/1\,\mathrm{ms^{-1}} = 4\,000\,\mathrm{m}$$

The total length is 400 000 meter, thus giving 100 boxes ($m = 100$), and with a total flow time of 400 000/3 000 seconds we get 134 time periods ($n = 134$), altogether m times $n = 13\,400$ iterations. With a PC-AT (80286 processor),

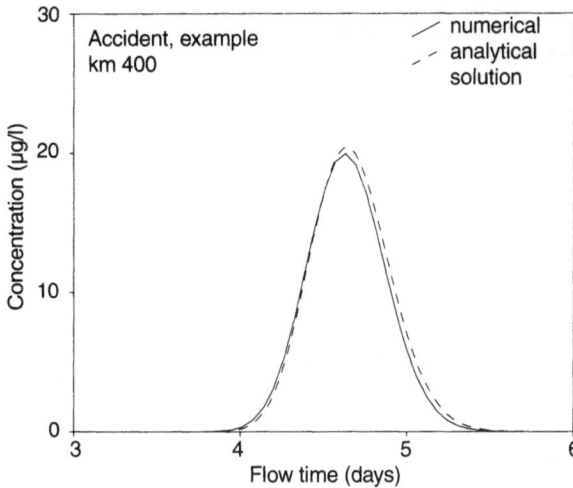

Fig. 3.5. Comparison of the analytical and numerical solutions for an accident on the Rhine

the calculation takes less than one minute. The comparison with the analytical solution (Fig. 3.5) demonstrates a good agreement.

The method is simple and can be used to simulate the fate of chemicals in rivers with varying hydrological properties. The example was used as an introduction to the problem of numerical solutions. Further information can be found in the literature, e.g. Vvedenski (1993), Strang and Fix (1973), Lapidus and Pinder (1982) and Richtmeyer and Morton (1967).

Example 3.5: 2, 4-Dichlorophenoxy acetic acid

On 21 November 1986, due to an accident, two tonnes of the herbicide 2,4-D (2,4-dichlorophenoxy acetic acid) were released into the Rhine (km 424, left side) over six hours (LAWA Landesamt für Wasser und Abfall Nordrhein-Westfalen 1987). In the European Union, the concentration of pesticides is not allowed to exceed 0.1μ g/L in drinking water. Following the accident, the use of bank-filtrated Rhine water for drinking water supply had to be stopped. 2,4-D was measured at Mainz (km 505), Bad Honnef (km 640) and Lobith (km 859), here by both German and the Dutch authorities. The results differ somewhat (IKSR 1987, Samenwerkende Rijn- en Maaswaterleidingbedrijven 1987).

Since relatively exact input data were available, the simulation was successful even with the easy model approach (Fig. 3.6). The example was taken from Matthies et al. (1992).

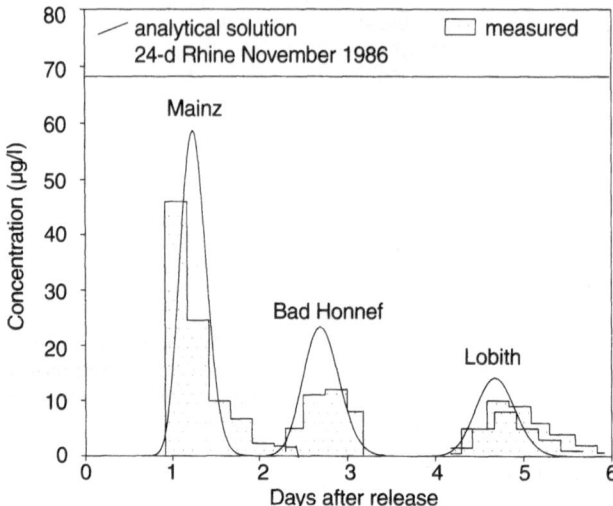

Fig. 3.6. Simulation of the 2,4-D accident

Exercises on Chapter Three

3.1 *Alarm!* On Sunday at 10:00 pm a tanker had an accident on the Rhine at km 430. One tonne of bromacil was released. What is the maximal concentration at Duisburg (km 780)?

Data of the Rhine: Width 333.3 m; depth 3 m; flow velocity 1 m/s; longitudinal dispersion D_L 500 m^{2}s^{-1}. No significant degradation.

3.2 On the following Monday at 10:00 a.m., a further 500 kg of Bromacil were released. What is the approximate maximum concentration at Duisburg?

3.3 Load CemoS, select the model SOIL. Enter the data of exercise 3.1 for *u* and D_L (units ... per day). Select the input function 'pulse input'. Calculate 3.1. Is this successful?

Chapter 4:
Partitioning of Chemicals in the Environment

4.1
Division into Phases

We all agree that our environment is manifold. For the application of chemical exposure models, simplifications need to be made. One way is to divide the environment into several phases.

Experiment One: Take a glass vessel and fill it just below half full with water. If you close the vessel, you have a mixture of air and water in it. As much as you try, you will not succeed in mixing both. Water and air are two different phases.

Experiment Two: Open the glass vessel and put an equal volume of alcohol (ethanol) into the water. The alcohol mixes easily with the water. Both form one phase.

Experiment Three: Add a few drops of olive oil to the water-alcohol phase, please don't shake. Despite the fact that olive oil has the same density as water with alcohol, the oil does not mix with the liquid, but forms beautiful spheres within the other phase.

Oil and water form separate phases that are not mixable (except, e.g. when using detergents or similar means). Even when the mixture is shaken, which makes it turn grey, the phases separate from each other again after a while.

Experiment Four (please be careful when you discard the petroleum): Try to mix petroleum with olive oil. You may have no problem here. Oil and petroleum form one phase.

In all experiments, a further phase that cannot be mixed is the glass itself. Conclusion: Several phases exist that cannot be mixed. In our experiments we have:

Hydrophilic phases (water and substances that are easily mixable or soluble in water, e.g. alcohol): These phases are polar.

Lipophilic phases (lipids, oils and compounds that mix with them): These phases are non-polar and hydrophobic.

Explanation:
Water H_2O has a dipole moment. The hydrogen atoms are partially positively charged and the oxygen atoms negatively. Electric interactions exist, the so-called hydrogen bonds. The relatively high boiling point of water compared to similar molecules, such as H_2S, and the density anomaly (lowest density at 4 °C) of water is also caused by the hydrogen bonds. This effect always occurs when the partially positively charged H-atom of an OH-, NH- or similar group comes close to an atom or molecule rich in electrons. On the other hand, very non-polar compounds distort this interaction, are pushed away and form their own phase.

Gas phases (air, gases, volatilized compounds): Gases have a different aggregate state. They are mixable with other gases.

Solid phases (glass, stones, rocks etc.): Solid phases have a different aggregate state. Solid phases can probably dissolve or evaporate, but as long as they do not, they are not mixable with the other phases.

4.2
The Partition Coefficient

Experiment Five: Add a droplet of methylene blue (or blue ink) to the water-alcohol-oil mix in the vessel. The water becomes blue, but the oil is not colored.

Explanation:
When a substance is dissolved in two neighbouring, non-mixable phases until saturation, the concentration ratio is in a certain, substance-dependent ratio. Even if the substance is not dissolved until saturation, a fixed concentration ratio occurs, presuming that the substance can mix freely in both phases. The ratio can be reproduced.

At this concentration ratio, the fugacities of the substance are equal in both phases (see Section 5.3); the system is in thermodynamic equilibrium. At small concentrations, the fugacity is directly proportional to the concentration of a substance (Mackay 1979, Mackay and Paterson 1981). Then the concentration ratio between both phases in equilibrium is independent of the absolute concentration.

The partition coefficient is deduced from thermodynamics (Nernst Partitioning Law) resp. from the chemical potential (Schwarzenbach et al. 1993).

The partition coefficient K is the concentration ratio in equilibrium:

$$K_{ij} = C_i/C_j \tag{4.1}$$

K is the partition coefficent (kg/m^3 to kg/m^3 or mol/m^3 to mol/m^3) and C is the equilibrium concentration (kg/m^3 or mol/m^3), i and j are the indices of the neighboring phases. In the literature, partition coefficients may also be called equilibrium constants.

Partitioning between air and water

Experiment Six: Open the vessel. You will smell alcohol. The alcohol obviously evaporates. However, it does not smell of fat. It seems as if some chemicals evaporate faster than others.

When concentrations of a liquid are small, the partitioning between air and water is described by Henry's Law (Atkins 1986). The partition coefficient, the so-called Henry's Law constant, can be calculated from solubility in water S and saturation vapor pressure p_s (Lyman et al. 1990):

$$H = p_s/S \qquad (4.2)$$

H is the Henry's law constant ($Pa\,m^3 mol^{-1}$), p_s is the saturation vapor pressure (Pa) (with solids: the saturation vapor pressure of the subcooled liquid) and S the solubility in water (here: $mol\,m^{-3}$). The partition coefficient air to water K_{AW} (= non-dimensional Henry's Law Constant) is then:

$$K_{AW} = H/(RT) = C_A/C_W \qquad (4.3)$$

C_A is the equilibrium concentration in air (kg/m^3), C_W is the equilibrium concentration in water (kg/m^3), R is the universal gas constant ($8.314\,J\,mol^{-1}\,K^{-1}$) and T is the temperature (K). Values of K_{AW} can vary by many orders of magnitude (Table 4.1). For estimation methods of K_{AW}, please refer to Chapter 11.

Table 4.1. Examples of the partition coefficients between air and water K_{AW} (Nirmalakhandan and Speece 1988)

Chemical	K_{AW}	Known contaminants	K_{AW}
n-Hexane	74.13	2,3,7,8-TCDD (Dioxin)	0.0015
n-Hexene	17.78	p,p'-DDT	$2.14 \cdot 10^{-3}$
Benzene	0.22	Atrazine	$8.05 \cdot 10^{-9}$
1,2-Dichloroethane	0.054	2,2',4,5,5'-PCB	0.0076
Methanol	$1.9 \cdot 10^{-4}$	Hexachlorobenzene (HCB)	0.054
Phenol	$1.6 \cdot 10^{-5}$	Trichloroethene	0.5
4-Bromophenol	$6.2 \cdot 10^{-6}$		
Bromacil	$3.65 \cdot 10^{-9}$		

Partitioning between lipid phases and water

The equilibrium partitioning between a hydrophobic phase (lipids, oils, etc.) and water is described by the *n-octanol-water partition coefficient* K_{OW} (unit kg substance per m^3 octanol to kg substance per m^3 water).

$$K_{OW} = C_O/C_W \qquad (4.4)$$

C_O is the equilibrium concentration of a substance in n-octanol, and C_W is that in water. The K_{OW} is used as a predictor for the partitioning between lipid phases in the environment and water (Mackay et al. 1985). Measured values are available for many compounds (Lyman et al. 1990, Suzuki and Kudo 1990, Rippen 1995). The K_{OW} can also vary by many orders of amount (Table 4.2).

N-octanol is an aliphatic, unchained compound with an OH-group at one end. It has similarities with fatty lipids. The partitioning between olive oil and water in former times was used (compare Experiment Four). However, olive oil is no pure and well defined substance, but may vary depending on origin, subspecies and environmental conditions. The determination of K_{OW}-values in a laboratory is achieved by partition experiments similar to Experiment Five. The substance is added to the octanol-water phases in the vessel, the vessel is shaken for some hours and the concentration in the water phase is determined. For very high values of K_{OW}, the determination is uncertain, since concentrations in water get small and sorption to the glass walls may disturb the results. This is why very high values of K_{OW} are sometimes uncertain. Modern methods use the HPLC retention times. For estimation methods of the K_{OW}, please refer to Chapter 11.

Table 4.2. Examples of K_{OW}-values for certain chemicals (Suzuki and Kudo 1990)

Chemical	K_{OW}	$\log K_{OW}$
Dimethylsulphoxide	0.0045	−2.347
Methanol	0.17	−0.77
Acetone	0.57	−0.24
Propanol	1.77	0.248
Hexanol	107.15	2.03
Toluene	537.03	2.73
Biphenyl	$1.23 \cdot 10^4$	4.09
2,4,5,2′,4′,5′- Hexachlorbiphenyl (PCB)	$5.24 \cdot 10^6$	6.72
Indeno[1,2,3-cd]pyrene	$4.57 \cdot 10^7$	7.66
Known contaminants		
2,3,7,8-TCDD	$5.75 \cdot 10^6$	6.76
Trichloroethene	195	2.29
Atrazine	512	2.71
Hexachlorbenzene (HCB)	$2.95 \cdot 10^5$	5.47

Sorption to soils and sediment

Sorption to solids is described by the empirical Freundlich relation (Tinsley 1979):

$$x/m_M = K C_W^{1/n} \qquad (4.5)$$

x is the adsorbed amount of chemical (g), m_M is the mass of sorbense, here the soil matrix M (g), K is the proportionality factor (Freundlich coefficient) (cm^3 water / g soil = g water / g soil), C_W is the equilibrium concentration in the aqueous solution (here: g/cm^3 water = g/g water) and n is a measure of non-linearity of the relation. For small concentrations, values of n are close to one (Tinsley 1979, Schwarzenbach et al. 1993). The Freundlich coefficient can then be seen as the slope of the linear adsorption/desorption isotherm. It is often called the distribution coefficient K_d between soil matrix and water.

$$x/m_M = C_M = K_d C_w \qquad (4.6)$$

C_M is the concentration sorbed to the soil matrix (g/g).

The linear sorption coefficient K_d can be determined by measuring the Freundlich adsorption isotherm (when exponent = 1) (Hamaker and Thompson 1972). Unpolar chemicals are mainly sorbed to the organic matter of the soil. K_d can therefore be estimated from the K_{OW} (via the K_{OC}, see below) and the organic carbon content. A correction factor Φ is used for the dissociation of acids and bases. No correction is necessary for measured K_d-values, please refer to Section 4.3.

Historically, the unit of the K_d is cm^3 water per g solid. The use of the unit g water per g solid avoids errors with wrong density units when the partition coefficient is calculated (unit mass per volume to mass per volume).

The natural bulk soil consists of soil matrix, soil solution and soil gas. The partition coefficient bulk soil to water K_{BW} is therefore

$$K_{BW} = C_B/C_W = K_d \rho_B/\rho_W + \theta + (\varepsilon - \theta)K_{AW} \qquad (4.7)$$

C_B is the equilibrium concentration in bulk soil (kg/m^3), C_W that in (external) water (kg/m^3), θ is the volumetric water fraction of the soil, ε is the volumetric total porosity, ρ_B is the dry soil density and ρ_W is the density of water. The last two terms on the right are normally not important for lipophilic chemicals.

Sediment consists only of matrix and pore water, the third term on the right is not necessary:

$$K_{SW} = C_S/C_W = K_d \rho_s/\rho_W + \theta \qquad (4.8)$$

S is the index for sediment.

The partition coefficient between *particles* and water is

$$K_{PaW} = K_d \rho_{Pa}/\rho_W \qquad (4.9)$$

Pa is the index for particles.

Estimation of the K_d-value

In soils and sediment, mainly humic substances form the organic carbon. They are partly of a hydrophobic nature and are mainly responsible for the sorption of lipophilic organic compounds. Therefore, lipophilic sorption is proportional to the organic carbon content (Karickhoff 1981):

$$K_d = K_{OC}OC \qquad (4.10)$$

K_{OC} is the partition coefficient between organic carbon and water and OC is the organic carbon content of the soil (g/g dry mass, also called C_{org}). K_{OC} is closely correlated to K_{OW}. A number of regressions was found (Lyman et al. 1990), among them:

$$K_{OC} = 0.411K_{OW} \qquad \text{(Karickhoff 1981)} \qquad (4.11a)$$

$$\log K_{OC} = 0.72 \log K_{OW} + 0.49 \qquad \text{(Schwarzenbach and Westall 1981)} \qquad (4.11b)$$

Beware!
These equations are only valid for non-dissociating organic compounds, and not for ions, dissociating acids, bases or amphoteres. For these substance classes, adsorption coefficients should be determined by other methods, e.g. experimentally, or eventually corrected (see Section 4.3). For extreme values of K_{OW}, i.e. small or large ones, there are considerable deviations between both equations.

Partition coefficient biota to water

Biota are a mixture of phases. It is often assumed that the sorption capacity of biological tissue, the so-called bioconcentration factor BCF, can be related to n-octanol by a log/log regression (refer to Chapter 11). This relation was often found for fish. However, a thermodynamical equilibrium partitioning is unlikely for predatory fish, since accumulation also occurs via the food intake and not only via the gills. Accumulation factors can also be defined for plant tissue, but they should be related to air and soil (refer to Chapters 9 and 11).

4.3
Partition Coefficients for Dissociating Chemicals

Acids and bases dissociate depending on the pH values of the solution. Acids are substances that form positively charged H^+ ions (better H_3O^+) in water. Bases form negatively charged hydroxyl ions OH^- (Hollemann-Wiberg 1976).

Acid reaction:

$$AH + H_2O \leftrightarrow H_3O^+ + A^-$$

Base reaction:

$$B + H_2O \leftrightarrow BH^+ + OH^-$$

$$BH^+ = \text{cation}, \quad A^- = \text{anion}.$$

The dissociation constant K_a is:

$$K_a = [A^-][H_3O^+]/[AH]$$

($[H_2O]$ drops out, constant)

Analogously K_b for bases

$$K_b = [BH^+][OH^-]/[B]$$

$[BH^+]$, $[A^-]$, $[AH]$, $[B]$, $[H_3O^+]$, $[OH^-]$ are the activities in mol/L (for small ion strength approximately identical to concentrations).
    Usually the dissociation constant is given as negative $\log_{10}$:

$$pK_a = -\log K_a$$

The pK_b-value of a base can be calculated from the pK_a-value of its conjugated acid:

$$pK_b = pK_W - pK_a$$

K_W is the ion product of water $= [H_3O^+][OH^-] = 10^{-14}$, $pK_W = 14$.
    The concentration of H_3O^+-ions in water is described by the pH:

$$pH = -\log[H_3O^+]$$

A pH of 7 shows 'neutral' reaction (activity of $[H^+] = [OH^-] = 10^{-7}$ mol/L). Values above 7 show alkaline reaction and values below 7 show acidic reaction.
    The dissociation (here of acids) at a given pH is calculated using the *Henderson-Hasselbalch* equation:

$$\log\{[A^-]/[HA]\} = pH - pK_a$$

The fraction of neutral species is then

$$\Phi = \frac{[HA]}{[HA] + [A^-]} = \frac{1}{1 + 10^{a(pH - pKa)}} \tag{4.12}$$

Φ is the fraction of neutral molecules, and therefore a correction factor in the case of acid-base reactions, $a = 1$ for acids and -1 for bases. For non-dissociating compounds, $\Phi = 1$.

Acid/base reactions have an influence on the partitioning. Ions are principally hydrophilic, because of the interaction of charged molecules with water. This implies that partition coefficients (for lipophilic interactions) only hold for the neutral fraction.

When measured values of K_d, K_{OC} or BCF are available, it has to be checked whether the pH is the same as in the scenario to be simulated. If not, the values need to be corrected. Therefore, first calculate the fraction of neutral molecules during measurement. Then determine the K_d, K_{OC} or BCF of the neutral species. Then calculate Φ, the fraction of neutral species at the simulated scenario, and multiply by K_d, K_{OC} or BCF of the neutral species. This procedure is then valid when the ion itself does not significantly sorb (e.g. to charged material such as clay). A correction for dissociation is necessary, too, when K_d, K_{OC} or BCF are calculated from K_{OW}. The procedure is:

$$K_d(pH, meas) \;=\; K_d(neutral) \cdot \Phi(meas)$$
$$\rightarrow K_d(neutral) \;=\; K_d(pH, meas)/\Phi(meas)$$
$$\rightarrow K_d(pH, sim) \;=\; K_d(neutral) \cdot \Phi(sim)$$

analogously for K_{OC}, BCF.

$K_d(pH, meas)$:   measured K_d at pH of measurement
$K_d(neutral)$: K_d of neutral species
$K_d(pH, sim)$:   calculated K_d at pH of simulated scenario
$\Phi(meas)$, $\Phi(sim)$:   fraction of neutral species at pH of measurement and simulated scenario.

The same correction is necessary when the K_{AW} has been calculated from solubility and partial pressure of the neutral substance, because only the neutral fraction evaporates:

$$K_{AW}(pH, sim) = K_{AW}(neutral) \cdot \Phi$$

$K_{AW}(pH, sim)$ is the partition coefficient atmosphere to water of the total substance at the pH of the simulated scenario, $K_{AW}(neutral)$ is that of the neutral species.

These corrections for ions are in particular necessary for those chemicals with pK_a-values that are environmentally relevant (3 to 9) (see Table 4.3).

In the following models, dissociation is always considered by the correction of K_d and K_{AW}. In CemoS, the correction is automatically done when the values are estimated.

No automatic correction is made for the BCF, because the pH in fish may differ from that in the surrounding medium.

For the important pollutant class of heavy metals that are usually ionized, sorption to soil or sediment matrix is not of a lipophilic nature. Sorption of cations to clay is more important and should be determined experimentally. Most heavy metals (except organometallic compounds) have no measurable vapor pressure, and $K_{AW} \rightarrow 0$.

Table 4.3. Dissociation constants (negative decadic logarithm, pK_a) (Rippen 1996)

Name	pka	Type	a
Phenol	9.9	Acid	1
2-chlorophenol	8.43	Acid	1
3-chlorophenol	8.92	Acid	1
4-chlorophenol	9.29	Acid	1
2,4-dichlorophenol	7.77	Acid	1
2,4,5-trichlorophenol	6.91	Acid	1
Pentachlorophenol	4.80	Acid	1
2,4-D	2.73	Acid	1
Aniline	4.61	Base	−1
Atrazine	1.68	Base	−1

4.4
Combination of Partition Coefficients

If the partition coefficients are of the same unit (here kg/m^3 to kg/m^3), then an unknown coefficient can be found from two known ones:

$$K_{il} = K_{ij}/K_{lj} = 1/K_{li} = C_i/C_l \qquad (4.13)$$

4.5
About Equilibrium

Air, water and hydrophobic phases that are not mixable with water are found in the biotic and abiotic environment. The calculation of equilibrium partition coefficients allows the estimation of the partition tendency of a chemical. This concept has been quite successful for the estimation of chemicals' fate. Together with diffusion and advection processes it is the basis of almost all exposure models.

The equilibrium is the condition with the highest entropy. This condition is only approximately reached within finite time periods. Diffusion processes always go towards higher entropy, i.e. to equilibrium. The smaller the scale, the closer the concentrations are at equilibrium (local equilibrium).

Exercise: Diffusion through phase boundaries

When diffusion through phase boundaries occurs, the partition coefficient plays a major role. At the boundary between two (in themselves) homogeneous phases there are laminar films on both sides. The substance transfer is controlled by diffusion through the films, Fig. 4.1 (two-film theory, Whitman 1923, Tinsley 1979, Schwarzenbach et al. 1993).

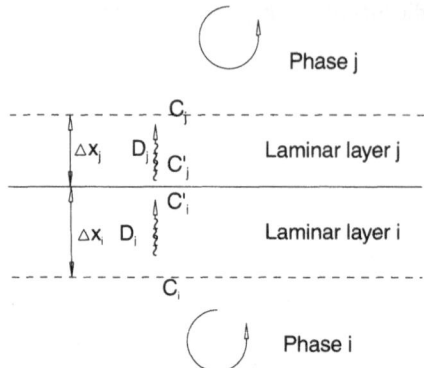

Fig. 4.1. Two-film model for exchange through phase boundaries

The diffusive net flux J_i per unit area $(kgs^{-1}m^{-2})$ in the laminar layer of phase i (and for phase i) is:

$$J_i = -D_i/\Delta x_i(C_i - C_i') = -g_i(C_i - C_i') \tag{4.14}$$

Concentrations at the boundary are in local equilibrium:

$$C_i'/C_j' = K_{ij}$$

The equation may be transformed to give

$$J_i = -D_i/\Delta x_i(C_i - C_j'K_{ij}) \tag{4.15}$$

The net flux through the film of phase j is

$$\begin{aligned} J_j &= -D_j/\Delta x_j(C_j' - C_j) = -g_j(C_j' - C_j) \\ &= -g_j(C_i'K_{ji} - C_j) = -g_jK_{ji}(C_i' - C_j/K_{ji}) \end{aligned} \tag{4.16}$$

The overall flux through the boundary per unit area is then

$$J_{ij} = -g_{ij}(C_i - C_j/K_{ji}) \tag{4.17}$$

With eqs. 3.3 and 3.4: $1/g_{ij} = 1/g_i + 1/(g_jK_{ji})$

g_{ij} is the overall conductance for diffusive exchange across the boundary between phases i and j (m/s).

Beware!
When the substance profit for phase j is calculated, the sign changes. The conductance still refers to diffusion in phase i, or g_{ij} is multiplied by the partition coefficient K_{ij}:

$$\begin{aligned} -g_{ij}(C_i - C_j/K_{ji}) &= g_{ij}(C_jK_{ij} - C_i) \\ &= g_{ij}K_{ij}(C_j - C_i/K_{ij}) \\ &= g_{ji}(C_j - C_i/K_{ij}) \end{aligned}$$

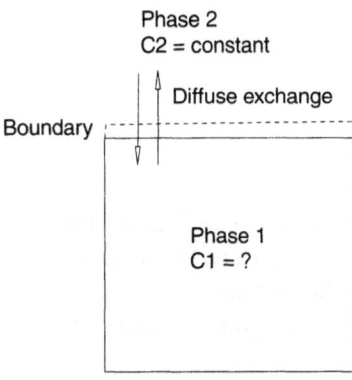

Fig. 4.2. Example of exchange across a phase boundary

The fact that the phase relation of the conductance value is not explicitly mentioned and cannot be seen at the unit, either, may cause confusion. But it can be seen from the formulation of the flux equation.

Example 4.1: Uptake in a compartment from its surroundings (Fig. 4.2)

Let us look at the diffusive flux between two compartments with concentrations C_1 and C_2. By definition, the mass in a compartment is homogeneously distributed, and there are no concentration gradients within the compartment, only between compartments. This is identical to the fact that mixing within the compartments is much faster than the exchange between them. The laminar films at the boundary may be seen as a 'semipermeable membrane'. We assume:

- Two phases 1 and 2 are neighboured by area A (m^2).
- Phase 2 is very large (e.g. atmosphere), therefore C_2 (mg/m^3) is quasi constant.
- The substance flux between the phases is controlled by diffusion in the boundary layers with the known overall conductance g_{21} (m/s).
- Each phase is in itself well mixed and homogeneous.
- The equilibrium condition is $K_{12} = C_1/C_2$.
- The initial concentration of C_1 is 0.

The mass balance equation of phase 1 is:

$$dm_1/dt = -A\,J_{12}$$

m_1 is the substance mass in phase 1, $m_1 = V_1 C_1$

$$
\begin{aligned}
V_1 dC_1/dt &= A g_{21}(C_2 - C_1/K_{12}) = \\
&= A g_{21} C_2 - A g_{21} C_1/K_{12} \\
&= -a C_1 + b
\end{aligned}
\tag{4.18}
$$

Question: What is the value of C_1 after 50 days, if $C_2 = 0.1\,mg/m^3$, $A = 1\,m^2$, $g_{21} = 0.001\,m/s$, $K_{12} = 10\,000$ and $V_1 = 0.1\,m^3$?

Analytical solution analogously to Example 2.1:

$$
\begin{aligned}
C_1(t) &= b/a(1 - e^{-at}) \\
&= C_2 K_{12}\{1 - \exp[-A g_{21} t/(V_1 K_{12})]\} \\
C_1(50\,\text{d}) &= 986.7\,\text{mg/m}^3
\end{aligned}
$$

In this example, the advantage of compartmentilization can immediately be seen. The spatial dependency of mass fluxes in the compartment is not explicitly considered. Mass flux is only through the boundary between compartments. The compartment model is a purely *kinetic* model; only time dependencies are considered.

Exercises on chapter four

4.1 The water of a river is slightly polluted, containing hexachlorbenzene (HCB) in a small concentration, $C_W = 1\,\mu\text{g/L}$ (= $10^{-6}\,\text{kg/m}^3$) = $1\,ppb$ (part per billion). The sediment of the slightly polluted river has 10% organic carbon. What is the equilibrium concentration C_S of HCB in the sediment (density $\rho_S = 2\,\text{g/cm}^3, \theta = $ pore water content = 50%)?

4.2 2,4-Dichlorphenoxy acetic acid (2,4-D) has a *pKa* of 2.73 and a log K_{OW} of 1.57. What is the K_d of the chemical to the above sediment at pH = 3, 5, 7?

4.3 Solve example 4.1 using the numerical Euler method.

Chapter 5:
Multi-Media Models

5.1
Introduction

In the 1960s and 1970s persistent organic chemicals were detected in almost all environmental media, often far away from their emission sources. The measurements of chloroorganic pesticides like DDT and Dieldrin could be explained by their dispersive use. However, other compounds like PCBs which are solely used in closed devices were also found ubiquitously. This observation led to the recognition that almost all chemicals have a tendency to be distributed over various environmental media of differing properties. The driving forces are advection and dispersion with wind and water and the partitioning between environmental phases.

The finding of the ubiquitous occurence of xenobiotics motivated the development of multi-media models which adequately describe chemicals environmental distribution and fate. Mackay and various co-workers (1979, 1981, 1985, 1992) published a series of articles on adapting the concept of fugacity from chemical engineering for the distribution of environmental chemicals. He also introduced the "Unit World" approach as a generic approach for a typical global environmental section. Klöpffer et al. (1982) at the same time developed a multi-media model which should explain the equilibrium partitioning between environmental media ('compartmentalization'). Since the beginning of the 1980s these approaches have been applied to further chemicals and other environmental sections.

In response to the need of quantitative methods for exposure prediction in chemical legislation multi-media models were introduced as evaluative models in national and supra-national regulations. For instance, the EU Technical Guidence Documents (EU 1996) recommend a multi-media model consisting of four compartments for estimating regional exposure levels in air, water, soil and sediment. In particular, new chemicals can be assessed by using only the base set of physico-chemical substance properties to describe their potential distribution and environmental fate. Measured concentrations of existing

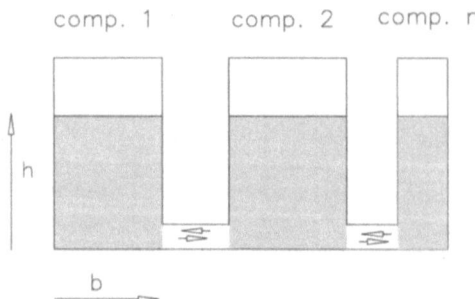

Fig. 5.1. Level 1; closed system in equilibrium; no source and sink

chemicals can be compared to calculated to get a better insight into the governing transport and transformation processes.

The description of the multi-media modeling approach follows the terminology of Mackay, first introduced in 1979 (Mackay and Paterson, 1979):

Level 1: Equilibrium, no reactions, closed system
Level 2: Equilibrium, open system, reactions, steady state
Level 3: Non-equlibrium, open system, reactions, steady-state
Level 4: Non-equilibrium, open system, reactions, non-steady state.

5.2
Partition Models for Multi-Compartment Systems

Level 1: Equilibrium, no reaction, closed system

It is assumed that the whole chemical mass in the system is distributed between the compartments according to the equilibrium partition coefficients. The substance mass is constant, there is no source and sink. The result is the partition tendency of a chemical in an environmental system. The principle is seen in Fig. 5.1. Compartments 1,2,...,n are shown as boxes. The substance mass (height multiplied by width) is indicated by the grey area. Free and immediate exchange is possible between the boxes. The height level in all boxes is the same, and the system is in equilibrium.

The total mass in the system m (kg) and volumes V_i (m^3) are given, concentrations C_i are unknown.

$$m = C_1 V_1 + C_2 V_2 + \ldots + C_n V_n \tag{5.1}$$

In equilibrium, we have $C_i/C_1 = K_{i1}$, $i = 1,\ldots,n$, and $K_{11} = 1$. It follows:

$$
\begin{aligned}
m &= C_1 V_1 + C_1 K_{21} V_2 + \ldots + C_1 K_{n1} V_n \\
C_1 &= \frac{m}{V_1 + K_{21} V_2 + \ldots + K_{n1} V_n} \\
C_i &= K_{i1} C_1; \quad m_i = V_i C_i
\end{aligned}
\tag{5.2}
$$

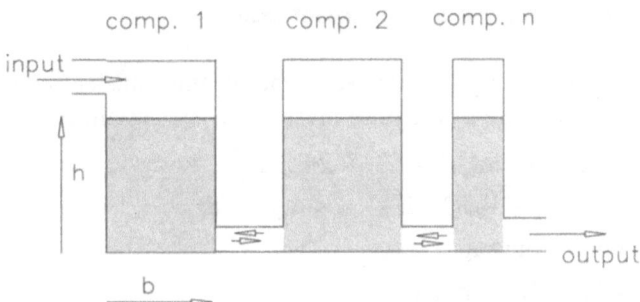

Fig. 5.2. Level 2; system in equilibrium, with source and sink

where $i = 1, \ldots, n = $ air, water, soil, sediment, fish ... this depends on the environmental system that is to be simulated.

Level 2: Equilibrium with source and sink, steady-state

Here we assume that a continuous input into the system exists. In steady-state, source and sink are balanced. Still we assume that thermodynamic equilibrium partitioning between the phases occurs. The system is shown in Fig. 5.2.

In steady-state $dm/dt = 0$ and total input (kg/s) = sum of all elimination processes

$$I = \sum_i (V_i C_i \lambda_i) \tag{5.3}$$

Using the equilibrium condition

$$C_1 = \frac{I}{\lambda_i V_1 + \lambda_2 K_{21} V_2 + \ldots + \lambda_n K_{n1} V_n} \tag{5.4}$$

(Note: Advection $Q \cdot C$ into the system is part of I; advection out of the system is defined as first order rate $\lambda = Q/V$, flux per volume).

The average elimination rate or the system's elimination rate λ_s for the chemical follows from

$$\begin{aligned}
I &= \sum_i (V_i C_i \lambda_i) \\
&= \lambda_s \sum_i (V_i C_i) \rightarrow \\
\lambda_s &= I / \sum_i (V_i C_i) = I/m \tag{5.5}
\end{aligned}$$

Level 2dyn: Equilibrium with source and sink, transient case

As before, equilibrium between the phases is assumed, but the transient concentration is calculated. The change of the total mass m_t in the system is

$$
\begin{aligned}
dm_t/dt &= dm_1/dt + dm_2/dt + \ldots + dm_n/dt \\
&= V_1 dC_1/dt + V_2 dC_2/dt + \ldots + V_n dC_n/dt \\
&= \text{input} - \text{output} \\
&= \sum_i I_i - \sum_i (V_i C_i \lambda_i)
\end{aligned}
$$

In equilibrium, we have $C_i/C_1 = K_{i1}$, $i = 1, \ldots, n$, and $K_{11} = 1$. It follows

$$
\begin{aligned}
& V_1 dC_1/dt + K_{21} V_2 dC_1/dt + \ldots + K_{n1} V_n dC_1/dt \\
& = \sum_i I_i - C_1 \sum_i (V_i K_{i1} \lambda_i)
\end{aligned}
$$

Now only C_1 remains unknown. The change with time dC_1/dt is

$$
dC_1/dt = \frac{\sum_i I_i - C_1 \sum_i (V_i K_{i1} \lambda_i)}{V_1 + K_{21} V_2 + \ldots + K_{n1} V_n} \tag{5.6}
$$

or

$$
dC_1/dt = -aC_1 + b
$$

$$
a = \frac{\sum_i (V_i K_{i1} \lambda_i)}{V_1 + K_{21} V_2 + \ldots + K_{n1} V_n}
$$

$$
b = \frac{\sum_i I_i}{V_1 + K_{21} V_2 + \ldots + K_{n1} V_n}
$$

The solution for $C_1(t)$ is

$$
C_1(t) = C_1(0) e^{-at} + b/a (1 - e^{-at})
$$

The steady-state solution for $t \to \infty$ follows

$$
C_1(\infty) = \frac{b}{a} = \frac{\sum_i I_i}{\sum_i (V_i K_{i1} \lambda_i)} \qquad i = 1, \ldots, n
$$

Concentrations C_2 to C_n are calculated from the equilibrium condition

$$
C_i = K_{i1} C_1 \qquad i = 2, \ldots, n
$$

Level 3: No equilibrium partitioning, sources and sinks, steady-state

Unlike level 2, we do not assume equilibrium between the compartments. For every single compartment, input and/or output may occur. The exchange between compartments is controlled by transfer resistances. Steady-state is assumed (Fig. 5.3).

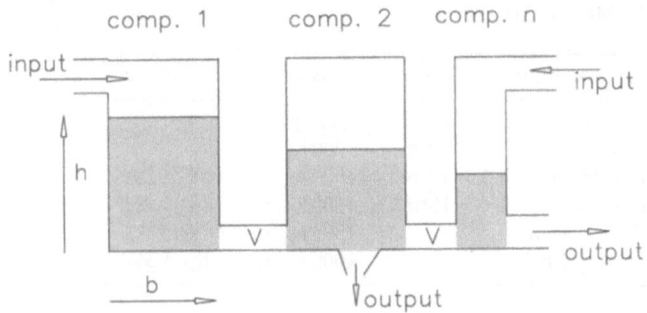

Fig. 5.3. Level 3; steady-state, system with input and output, no equilibrium

In the environment, two exchange processes exist: diffusion and advection.

1. Diffusive (compare Sect. 4.5)

$$N_{ij} = -Ag_{ij}(C_i - C_j/K_{ji}) = -N_{ji}$$

2. Advective

$$N_i = Q_i[C_i(\text{in}) - C_i(\text{out})]$$

Q_i is the flux of the medium (m^3/s).

Change of substance mass in the compartment =
+ input
± advection
± profit/loss from other compartments
− degradation
= 0 (steady-state)

$$V_i dC_i/dt = I_i + N_i + \sum_j (N_{ij}) - C_i V_i \lambda_i = 0 \tag{5.7}$$

for $i = 1,\dots,n$ and $j = 1,\dots,n$ and $j \neq i$

This gives an equation system of n linear equations with n unknown C_i. It is solved using the usual methods, e.g. by Gaussian elimination.

Level 4: No equilibrium partitioning, sources and sinks, transient case

Only in exceptional cases is really a steady-state in the environment. New to level 3 is now the transient simulation of mass balance and mass fluxes.

Change of substance mass in the compartment =
+ input

Table 5.1. Unit World of D. Mackay (1991)

compartment	volume	density	other
air	$1\,km^2 \times 6\,km = 6 \times 10^9 m^3$	$1.19\,kg/m^3$	
water	$0.7\,km^2 \times 10\,m = 7 \times 10^6 m^3$	$1000\,kg/m^3$	
soil	$0.3\,km^2 \times 15\,cm = 4.5 \times 10^4 m^3$	$1500\,kg/m^3$	OC 2%
sediment	$0.7\,km^2 \times 3\,cm = 2.1 \times 10^4 m^3$	$1500\,kg/m^3$	OC 4%
particles	$35\,m^3$	$1500\,kg/m^3$	OC 4%
biota (fish)	$7\,m^3$	$1000\,kg/m^3$	lipid 5%

$\pm$ advection
$\pm$ profit/loss from other compartments
$-$ degradation
$\neq 0$ (no steady-state)

$$V_i dC_i/dt = I_i + N_i + \sum_j (N_{ij}) - C_i V_i \lambda_i \neq 0 \qquad (5.8)$$

The solution can either be carried out analytically (in the case of constant or pulse input) or by numerical integration routines, e.g. Runge-Kutta or Euler (Chapter 2).

Now, initial conditions need to be known. V_i, Q_i, λ_i, I_i and other parameters can vary during the simulation. This makes the model applicable to a larger range of problems.

5.3
'Mackay Models'

In 1979, D. Mackay presented his *Unit World* of 1 km² area and proposed fugacity based partition models. At the same time, Walter Klöpffer (Klöpffer et al. 1982) from the Batelle Institute in Frankfurt/Main proposed a similar model.

The unit world consists of water (depth 10 m, 70% of area), soil (depth 15 cm, 30% of area), air (height 6000 m), sediment (depth 3 cm, 70% of area), particles (5 g/m³ water) and biota (fish, 1 g/m³ water); Mackay is still working on this concept, see details in Mackay (1991).

The fugacity approach

First note that all calculations that can be made with the fugacity concept can also be made using the 'traditional' concept described above, and vice versa. Both approaches differ only in the mathematical formulation.

Fugacity is a thermodynamic property. It is equal to the partial pressure of a substance in a compartment. The basic equation is

$$C = fZ \qquad (5.9)$$

C is the concentration of a substance (here: mol/m^3), f is the fugacity (Pascal, Pa = N/m^2) and Z is the capacity (mol m^{-3} Pa^{-1}). In equilibrium, fugacities f_i and f_j of two phases are equal, and thus

$$C_i/C_j = \frac{f_i Z_i}{f_j Z_j} = K_{ij} \qquad (5.10)$$

In Figs. 5.1 to 5.3, fugacity is height h of the grey area, Z is width b and concentration is the grey area $h \cdot b$. In equilibrium, fugacity f is equal in all phases, but not concentration C, since C depends on capacity Z.

The fugacity concept has formal similarities to the theory of heat: The heat capacity is analogous to the fugacity capacity, and the temperature to fugacity. Two phases are in equilibrium when the temperature (the fugacity) is equal.

Capacity of air Z_A

The fugacity of air can be deduced from the general equation of ideal gases:

$$pV = nRT$$

p is the partial pressure of the substance (Pa), V the volume (m^3), n the amount of substance (mol), R the universal gas constant (8.314 J mol^{-1} K^{-1}) and T the absolute temperature (K).

For real gases, partial pressure p is replaced by fugacity f (unit Pascal), then $fV = nRT$.

By rearrangement, it follows

$$n/V = f/(RT) \qquad \text{compare to}$$

$$C = fZ$$

with $C = n/V$ and f = fugacity, Z_A(mol m^{-3}Pa^{-1}) is:

$$Z_A = 1/(RT) \qquad (5.11)$$

Capacity of water Z_W

By definition (Eq. 5.9),

$$K_{AW} = Z_A/Z_W \qquad \text{rearranged and with Eq. 4.3}$$

$$Z_W = Z_A/K_{AW} = [1/(RT)]/[H/(RT)] = 1/H \qquad (5.12)$$

H is Henry's Law Constant (unit: Pa m^3/mol).

Capacity of sorbing phases

Z values of compartments with sorbing phases (e.g. soil, biota, sediment) are calculated from the partition coefficients to water K_{XW} by dividing them by $1/H$, or by multiplying them by Z_W:

$$Z_X = K_{XW} Z_W = K_{XW}/H \tag{5.13}$$

Diffusive transport processes with the fugacity concept

The diffusive net flux N_{ij} between phase i and j is driven by the fugacity difference:

$$N_{ij} = D_{ij}(f_i - f_j) \tag{5.14}$$

D_{ij} is an exchange term (unit: mol s^{-1} Pa^{-1}) and not the diffusion coefficient. The partition coefficient does not occur in the flux equation; but when D_{ij} is calculated, Z values are needed. Diffusive transport processes are well described by the fugacity concept, because net diffusion always goes from higher to lower fugacity (compare Sect. 4.5). See Mackay et al. 1985 or 1991 for the determination of D_{ij}-values.

Mass fluxes by advection are calculated by using fugacity:

$$N_i = Q_i f_i Z_i \quad (= Q_i C_i) \tag{5.15}$$

The fugacity concept does not necessarily ease multi-media modeling. The question whether or not fugacities are preferred depends on the user. In practice, concentrations are measured.

We do not use the fugacity concept for our modeling. But it was and is very popular, because Mackay proposed a common concept for all compartments of the environment. So now here are the four levels as fugacity models (Mackay 1979, 1981), known as 'Mackay models level 1 to 4':

Level 1: Equilibrium, no reaction, closed system

$$f_i = m / \sum_i (V_i Z_i) \tag{5.16}$$

and

$$
\begin{aligned}
C_i &= f_i Z_i \quad \text{and} \quad m_i = f_i Z_i V_i \\
f_1 &= f_2 = \ldots = f_n \quad \text{(equilibrium condition)}
\end{aligned}
$$

Level 2: Equilibrium with source and sink, steady-state

$$f_i = \frac{I}{\sum_i (V_i \lambda_i Z_i)} \tag{5.17}$$

$$f_1 = f_2 = \ldots = f_n \quad \text{(equilibrium condition)}$$

Level 3: No equilibrium, sources and sinks, steady-state

$$I_i - V_i f_i Z_i \lambda_i - \sum_j D_{ij}(f_i - f_j) = 0 \tag{5.18}$$

$$f_1 \neq f_2 \neq \ldots \neq f_n \quad \text{(no equilibrium)}$$

Level 4: No equilibrium, sources and sinks, transient case

$$V_i Z_i (df_i/dt) = I_i(t) - f_i(V_i Z_i \lambda_i + \sum_j D_{ij}) + \sum_j (D_{ij} f_j) \tag{5.19}$$

$$f_1 \neq f_2 \neq \ldots \neq f_n \quad \text{(no equilibrium)}$$

5.4
What Do These Models Tell Us?

Level 1: Equilibrium, no reaction, closed system

The complete global equilibrium will never be reached (maximal entropy). Generally, it is wrong to assume equilibrium on a larger scale. But all diffusive fluxes are directed towards the equilibrium. For smaller scales, this assumption may be very useful and is often used within more complex models (local equilibrium assumption). For larger scales, the assumption might be approximately valid for compounds that have been present for many years, are distributed widely (ubiquitously) and are relatively persistent.

Level 2: equilibrium with source and sink, steady-state

This type of model is subject to the same limitations as the one above. But it may serve for the calculation of mass balances of larger environmental systems with small data requirements. Such questions can be addressed as: Does the fast degradation rate in one compartment really mean significant degradation in the system? Background concentrations can be calculated with an accuracy of about an order of magnitude (sometimes more, sometimes less). The time to 95% steady-state can be found.

Level 3: No equilibrium partitioning, sources and sinks, steady-state

This type is advantageous, because the input compartment can be defined. Sometimes, however, it is extremely difficult to estimate exchange rates (e.g. soil to air). The data requirement is far greater than before. In principle, the same questions as with a level 2 model are addressed. A Mackay level 3 model (Mackay et al. 1992) has been proposed by the OECD as a reference model for the comparison of chemicals on a regional scale (alongside SAMS as a local model system, Matthies et al. 1992a).

Still some criticism is justified: Will near steady-state be reached? and when? How accurate are the simulations? How certain are the input data (in particular, transfer and reaction rates)? What is the real exchange between environmental media?

Due to the transfer rates, level 3 needs much more input data than level 2, and a few different parameters can significantly change result. The uncertainty is generally higher.

Level 4: Dynamic model

Real environmental processes are always dynamic. But for real environmental systems, the Mackay type models may be too global. Specially developed models are often superior.

Summary

With the levels 1–4 models, predictions of potential accumulation, environmental persistence and mobility can be made. The necessary data requirement is small and more or less available. Experience made with a reference substance can be used to enhance the accuracy of the next simulated chemical. This allows a ranking of chemicals. In this context, and as a screening and priority setting tool, these models are very helpful.

5.5
Regional Exposure Model

Local, regional and global models

An environmental fate model describes the dynamics of the transport and transformation of a chemical in a *specific* environment taking into account the physical, chemical, and biological processes which affect the quantity, structure, concentration and properties of the chemical on the considered spatial-temporal scale. It is a "mathematical converter" which transforms input rates (Mass/Time) into concentrations (Mass/Volume) under specific assumptions, initial and boundary conditions. Two approaches were mainly used to solve the problem:

– small scale approach, called the *local fate model* and
– large scale approach, called the *global fate model*

Local fate models are those which are concerned with the environment close to the release site (OECD 1992). The dominant processes are physical, i.e. advection and dispersion, mixing and dilution. A local model is usually a single medium model. Most of the models described in this book are local models, e.g. PLUME (Gaussian plume model, Chapter Eight). *Global* fate models are those which are concerned with large scale behavior (OECD 1992). Here the chemical equilibrium and transformation reactions are much more important, e.g. partitioning, degradation and diffusive mass transfer. A multi-media model is the most appropriate. Both approaches are simplifications by neglecting processes and environmental factors. Only the most sensitive parameters and processes are considered thus leading to a reduced set of equations and data.

A *regional* model focuses on the scale in between these two extreme scales. Thus, all environmental processes and factors have to be taken into consideration. This is obviously the most complex problem which has only recently been looked at. The description of chemicals regional fate poses difficult quantitative problems. Many different variables may be required which are often not available. This situation results in what is called the dilemma of middle number systems (Allen and Starr 1982, Turner and Gardner, 1991).

Why regional fate modeling?

The development of regional fate models is motivated by several reasons:

1. Chemicals are often diffusively released because of their wide-spread use. Specific sources can then not be identified (benzene from gasoline stations, spraying of pesticides, evaporation of solvents, or volatilization of plasticizers).
2. Chemicals are often emitted from a large number of point sources, e.g. stacks, vents or municipal waste water treatment plants (e.g. > 9000 in Germany, UBA 1995).
3. Persistent chemicals can be distributed over large areas up to a continental or global scale.
4. The regional boundaries and characteristics govern the transport and fate (a watershed, wind velocity, soil properties, land-use, temperature etc).
5. Import from and export into the neighboured region or country determine the residence time, mass balance and concentrations.
6. Environmental fate models can only be validated by comparing observed with simulated concentration when the appropriate environmental systems properties are well defined and known.
7. Environmental risk assessment of new and existing chemicals involves the prediction of environmental concentrations (PEC) on a regional level (*regional PECs*) besides local PECs (EU 1995, Ahlers et al. 1994).
8. Pesticides use should be regulated taking into account the different regional characteristics (Lepper et al. 1994).

How to define a regional exposure model?

The problem in defining regional fate models can be tackled from four major point of views. They represent various, but overlapping methods developed by different disciplines:

1. Natural landscape and land-use structure (Regional geography and land-scape ecology)
2. Socio-economic structure (Chemical economy and social-economic geography)
3. Substance specific properties (Environmental chemistry)
4. System integration level and system boundaries (Environmental systems analysis and fate modeling)

All four major items determine the definition of the regional spatial-temporal scale appropriate for the fate modeling purpose. The various disciplines have developed their own methods which can be checked for their utility.

Natural landscape and land-use structure (Regional geography and landscape ecology)

A *region* is an area with definite boundaries and characteristics. In geographical terms, it is a specific level of spatial and temporal structure and organization which is determined by the following aspects (Haase 1991):

– A region is characterized by its topographic structure (macro relief) repre-senting the geological evolution.
– A region has a pattern of soil types originating from the soil genesis.
– A region has a distinct macro climate which determines the ecophysiological effective energy (heat) and water balance including the formation of fresh water systems, e.g. rivers and lakes.
– A region covers a typical formation of vegetation with biogeochemical matter circulation.
– A region has its own land-use structure (agriculture, forest, cities, roads, waterways etc.) and population density and distribution.

Region is often used synonymously with landscape although both terms are not identical. Regional geography emphasizes the abiotic factors and landscape ecology the biotic factors. A hierarchy of spatial scales from nano (10^{-1} km^2) to macro (10^3 km^2) and global scale has been defined in geography. The typical regional scale is 10^4 to 10^6 km^2.

A region and its properties can be determined by using the maps and data sets for the above mentioned characteristics. Preferably, they should be given in the same cartographic scale. Vector as well as raster maps can be used. The maps and data sets are best stored and used in a *Geographical Information System (GIS)* by which large amounts of information can effectively be handled. The actual land-use structure can be derived from national statistics or satellite images. Climate zones were proposed by Walther and Lieth (1967). Regions can then be defined by combining the layers of the various spatial databases taking into consideration the appropriate scale. The result is a pattern of regions based on the natural and cultural landscapes.

Economic and demographic structure

Environmental concentrations of chemicals are determined by the economic and demographic structure of a region. For instance, Calamari et al. (1994) showed that the fingerprint of persistent chloroorganic compounds in vegetation is dominated by the socio-economics of a region or a country and of the past and present use of industrial chemicals and pesticides. Baccini and Brunner (1992) developed a method for assessing the regional material flows and balances.

The spatial and temporal pattern of chemical releases is determined by the economy operating in that region and the population living there. Release estimations are based on the whole life cycle of a chemical (OECD 1992). The locations of point sources and their characteristics, e.g. stacks or WWTPs, continuous or single releases, should be known. Diffuse area releases can be estimated from the use pattern and use categories (OECD 1992). The national economy is usually classified into distinct classes. Their economics as well as material balances are collected by the national bureaus of statistics. A list of pollutants emitted from these industrial branches was derived for Germany and is available from the UBA. The data can be processed in form of a matrix. Those industrial branches which are located in the region, e.g. along a river, can be stored in a GIS together with the data on their specific discharges. Regional specific releases can then be estimated for the region under consideration.

Substance Properties

Once a chemical is released from the technosphere it is out of control by man. The environmental forces like wind velocity, water flow, sunlight and microbial activity etc. interact with the physical and chemical properties of the substance and in this way determine and control its transport and fate. This action of the environment can be viewed as the "functional" processes of the environmental system. The subsystems soil, air, water, plants, animals, and human beings form the spatial structure in which the processes operate. The processes can be classified into physical transport, chemical reactions, and biological actions. Physical and biological processes determine the mobility and accumulation, whereas persistence is governed by all processes.

Degradation is the decomposition of the chemical structure of a substance. The chemical processes causing degradation of xenobiotics are called "sinks". The persistence of a substance is the result of either the absence or inefficiency of sinks or of the inability to reach potential sinks (Klöpffer, 1994). The key property of the substance is therefore its instability against the destructive environmental forces. Persistent chemicals can accumulate in biota or soil thus remaining in the environment over long periods. Junge showed, as early as 1975, that the variablity of measured air concentrations is reciprocal to the persistence of a chemical. A non-persistent compound shows larger concentration ranges than a persistent one. The residence time of a substance in a region depends on all dynamic processes:

1. Partitioning determines the mobility and accumulation of a chemical.
2. Degradability determines the potential of being decomposed.
3. Mass transfer rates determine whether a compound can enter a compartment where it could be degraded.

Thus, the appropriate scale for environmental transport and fate modeling depends on the physical and chemical properties of the substance itself.

Regional multi-media fate model

Mackay et al.(1992) applied a top-down approach to define a generic regional multi-media fate model. The evaluative or generic environment treated is an area of $100\,000\,\text{km}^2$ which is about the area of Greece. The SETAC-Task Force on the application of multi-media models for regulatory purposes (1994) proposed an area of $50\,000\,\text{km}^2$. The landscape structure in Mackay's model is represented by 90% homogeneous soil and 10% water. In other multi-media models soil is divided into three different land-use classes (arable, urban and natural soil). Multi-media regional fate models have been developed for some countries (e.g. Canada, France, The Netherlands). The geographical characteristics of generic regional environments have been specified for various European countries in ECETOC (1994).

Multi-media models of the Mackay type are mass-balance models with zero dimension, i.e. the spatial pattern is not explicitly considered. The compartments are all well mixed and have no internal structure. For air and water discharges the distance from a point source where complete mixing is achieved can be estimated taking into account the dispersion coefficients. Soil is itself heterogeneous and the assumption of homogeneous distribution is not valid for soil, at least for adsorbing compounds.

In the Technical Guidance Documents for chemical regulation within the EU a multi-media model of the Mackay type Level 3 was defined taking into account air, soil, surface water and groundwater. The fate of a chemical released into a distinct country or region can be calculated by using national data sets on area, climate, soil, rivers etc. Furthermore, a European generic data set should represent the typical situation when other data are not available.

The model is described in van Leeuwen and Hermens (1995) and was implemented in the HAZCHEM and EUSES computer code.

Example 5.1

The partition models may be used for the simulation of mass balances of regions. We show this by simulating 2,3,7,8-TCDD ('Seveso-dioxin') and benzene for Germany (data from Rippen 1996).

Both substances are rather different: Dioxin is extremely toxic, but is emitted in very small amounts. It is persistent and accumulates. Benzene is emitted in very high amounts (mainly from fuel), but it is not stable in the environment. It is a significant atmospheric pollutant and cancerogenic (LAI 1992).

Simulation of benzene

Due to the interim report of the German Enquete Commission "Schutz des Menschen und seiner Umwelt – Bewertungskriterien und Perspektiven für umweltverträgliche Stoffkreisläufe in der Industriegesellschaft" (Enquete 1993), 56 100 000 kg of benzene were emitted in Germany in 1991. Benzene degrades in air with a half-life of approximately 14 days, and in soil with a half-life of approximately 12 days (Rippen 1996). The equilibrium concentration ratio is calculated using the equations in Chapter Four. Only air and soil are considered.

Benzene data

Degradation half-life in soil 12 days; degradation half-life in air 14 days; Input 56 100 tonnes per annum; $M = 78$g/mol; $\log K_{OW} = 2.13$; $K_{AW} = 0.225$; $K_{BA} = 8.89$; for further values see Table 5.1, area is Germany in 1991.

We first consider the concentration course in air and soil until a quasi-steady-state is reached, then the decrease after a total cessation of emissions.

The result of the simulation is shown in Fig. 5.4. Due to its high mobility and relatively fast degradation in the environment, benzene is close to steady-state within a few months. The calculated concentrations are $1.5\,\mu g/m^3$ in air and $7\,ng/kg$ in soil. Measured concentrations of benzene in the air of rural German areas are below $1\,\mu g/m^3$, whereas in urban areas with high traffic, up to $30\,\mu g/m^3$ were found. In soil, concentrations of benzene are below $60\,ng/kg$ (Enquete 1993).

In the steady-state, 3 100 tonnes of benzene are in the air, but only about 0.7 tonnes are in the soil. Benzene is an air pollutant. It has no high accumulation potential, but is very mobile, due to the large amounts in the air. If all emissions

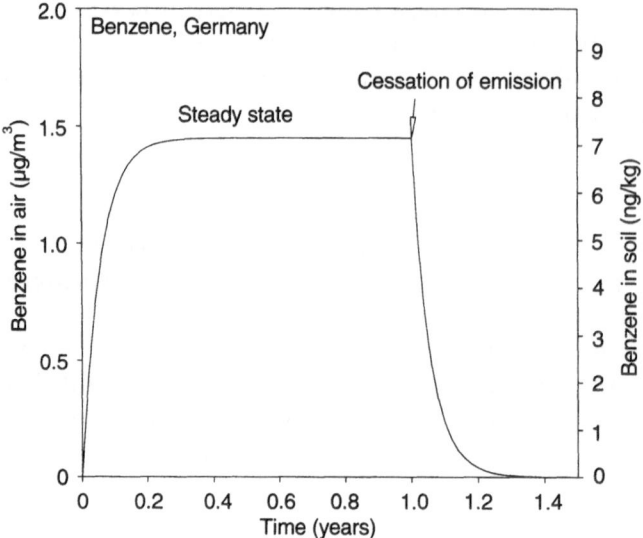

Fig. 5.4. Simulation of benzene with Level 2dyn, Germany, air and soil

of benzene were stopped (in neighboring countries, too), the concentrations of benzene in the environment would sink to close to zero within a few months.

Simulation of 2,3,7,8-TCDD

The environmental fate of the so-called 'Seveso dioxin' 2,3,7,8-TCDD is different to that of benzene. It is emitted in very small amounts, estimates for 1990 are 0.1 kg per annum (former West Germany). In the atmosphere, 2,3,7,8-TCDD is degraded by photolysis with a half-life of approximately 32 days. It is very persistent in soil. It is not clear whether a significant degradation occurs at all. We assume a half-life of 60 000 days or 164 years (Rippen 1993). The result of the simulation of the behavior of 2,3,7,8-TCDD in the soil-air system with the Level 2 model is shown in Fig. 5.5

Data 2,3,7,8-TCDD

Degradation half-life in soil 60 000 days; degradation half-life in the atmosphere 32 days; input is 0.1 kg/a; $M = 321.97$g/mol; $\log K_{OW} = 6.76$; $K_{AW} = 0.0015$; $K_{BA} = 0.47 \cdot 10^8$;

Scenario former West Germany: $V_{air} = 0.15 \cdot 10^{16}$m^3; $V_{soil} = 0.375 \cdot 10^{11}$m^3, other further values see Table 5.1.

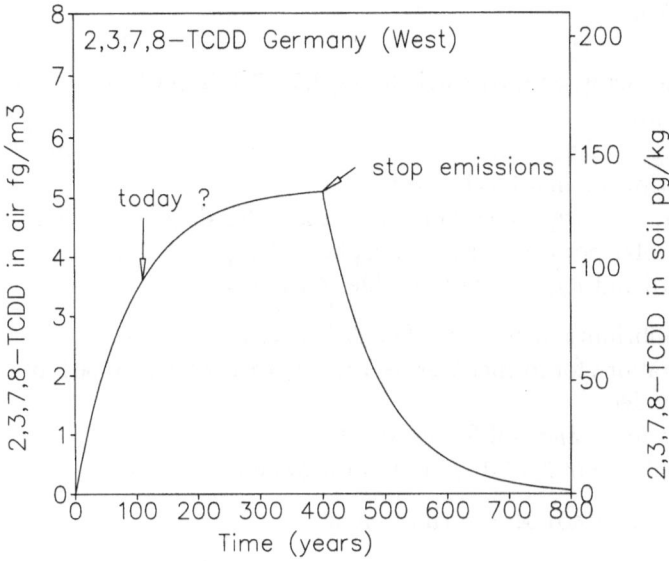

Fig. 5.5. Simulation of 2,3,7,8-TCDD with Level 2dyn, Germany

Results for 2,3,7,8-TCDD Level 2, steady-state

Amount in soil 9.2 kg, amount in air 0.008 kg, $C_{soil} = 135.9$ pg/kg fresh weight, $C_{air} = 5.17$ fg/m^3, half-life in the system is 63.6 years, steady-state (95%) after 257 years.

Comparison to measured values (McLachlan 1992): 2,3,7,8-TCDD in soil 70 pg/kg, in air 3.6 fg/m^3; simulated values are somewhat below steady-state.

2,3,7,8-TCDD has a very high potential for accumulation (in fish and plants as well as in milk and meat). It is very persistent in the environment, but its mobility is small.

Limits of the model

Due to the rigid assumptions, the accuracy of the model and the predictive power is limited. The calculated concentrations are larger than measured values in rural areas and smaller than those close to the emittents. They only hold for the lower troposphere (6 000 m height) and the top soil (15 cm depth). Other models are available for the calculation of concentrations close to the emitters and under special meteorological conditions, e.g. plume models (Chapter Eight). However, single emissions can only be considered for small areas. For regions with hundreds or thousands of emitters, and for chemicals emitted from area sources (benzene: traffic), the compartment approach might be useful.

Exercises on Chapter Five

5.1 In former West Germany, approximately 0.1 kg 2,3,7,8-TCDD (dioxin) were emitted per year (estimate).

TCDD in air is degraded by photolysis with a half-life of 32 days. In soil, it is very persistent (estimated half-life 160 years).

Data: soil depth 15 cm; organic carbon content 2%; dry density 1.5 g/cm^3; water content 30%, total porosity 50%; atmospheric height 6 000 m; 2,3,7,8-TCDD: log $K_{OW} = 6.76$ and $K_{AW} = 0.0015$ (Tables 4.1 and 4.2).

a) Calculate the equilibrium partition coefficient between air and soil.
b) Calculate concentrations for former West Germany (area about 250 000 km^2) with the level 2 model.
c) Calculate amounts in air and soil for steady-state.
d) How long is the half-life of 2,3,7,8- TCDD in the system?

Note: You can use the file EX51.MIF to run CemoS.

5.2 You rent a very nice room with wood paneling in an old house. You have some wood analyzed in our laboratory for wood preserving chemicals. A shock: 100 mg/kg Lindane (among other chemicals...).

a) Calculate concentrations in the room air. Assume equilibrium conditions.
   Data: Volume of wood 0.1 m^3, weight 50 kg. Room area 20 m^2, height 2.5 m; partition coefficient between wood and water: 1 000 (estimate); partition coefficient between air and water K_{AW} : 10^{-4}.
b) How much Lindane do you inhale in one hour (1 m^3 air)? Which other uptake pathways might occur?

5.3 You open the window. Per minute, you exchange 1 m^3 air. How long do you need to open the window, until less than 1% of the initial amount of Lindane remains in the room?

5.4 Run CemoS, load the substance hexachlorobenzene, select the LEVEL2 model and calculate the concentrations with the standard scenario (Germany).

Input data: initial amount 0.0 kg; input 1 000 kg/a, $C_{in} = 0.0$ kg/m^3; degradation rate constant in sediment = that in soil; degradation rate constant in fish = 0;

a) What are the steady-state concentrations? Where are the lowest and highest concentrations?
b) How much HCB is there in the compartments?

Chapter 6:
Contaminants in Surface Water

6.1
Introduction

Surface water is very often loaded with chemicals. Waste water, whether treated or not, is nearly always released into rivers, lakes or coastal waters.

EXPERIMENT: Take a glass of water, add some soil to it and stir it. The water becomes cloudy from the particles, which then slowly settle. These processes are similar to those found in rivers. There too, particles are suspended and sedimented. Rivers consist of water, sediment and biota, and exchange with the atmosphere occurs.

6.2
Analytical Steady-State Surface Water Model WATER

The steady-state box model WATER was developed to compare chemicals' behavior after being released into a river or lake. It includes the typical processes but in an idealized manner. WATER is part of the CemoS package. It is similar to RIVER, which was programmed for the OECD as part of SAMS (Screening Assessment Model System, Matthies et al. 1992a). WATER requires data that are available from the Screening Information Data Set (e.g. K_{OW}, K_{AW}, biodegradation rates). It gives estimates of concentrations, mass balances and flows.

When the concentration and fate of chemicals after continuous release into rivers are considered, then the assumption of steady-state simplifies the problem. Further model simplification is achieved by assuming homogeneous concentrations in a one-dimensional water body (fluid). Then plume effects and dispersion need not be considered. Remaining processes are

– dilution
– advective transport
– protolysis (dissociation of acids and bases)
– sorption to suspended particles

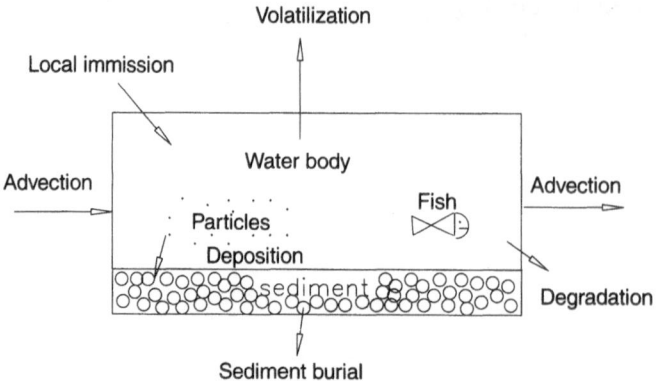

Fig. 6.1. A river segment in the WATER model

– net deposition of particle-bound substances
– degradation processes (sum of hydrolysis, photolysis, biodegradation and others)
– volatilization

The river is considered as being composed of a fluid compartment which describes the flowing water including suspensions and biota (fish) within it. It is considered as having exchanges with sediment and air (Fig. 6.1).

Initial concentration and transport equation

The average initial concentration C_0 follows from the dilution of the emitted waste water. WATER describes the dilution of a released chemical by assuming immediate mixing. When the average waste water flow and river flow Q_E and Q are given, the dilution factor Df is:

$$Df = Q/Q_E \tag{6.1}$$

The average concentration in the river close to the point of release, C_0 (kg/m³) is calculated from the concentration in waste water, C_E (kg/m³), river flow Q (m³/s) and waste water flow Q_E (m³/s):

$$C_0 = C_E Q_E/Q = C_E/Df \tag{6.2}$$

All elimination processes during advective transport are approximated by a first order rate process:

$$dC/dt = -\lambda C \tag{6.3}$$

λ : sum of elimination rates (first order, 1/time)
C : total concentration of a chemical in fluid (kg/m³)
t : time

The concentration is a function of distance $\Delta x = x - x_0$, initial concentration C_0 and time $\Delta t = t - t_0$. If the flow velocity is taken as constant for the whole segment, then:

$$\Delta t = \Delta x / u \qquad (6.4)$$

and

$$dC/dx = -\lambda/uC \qquad (6.5)$$

The integration of Eq. 6.5 with $C = C_0$ at $x_0 = 0$ and $t_0 = 0$ gives the analytical solution

$$C(x) = C_0 e^{-\lambda x/u} \qquad (6.6)$$

The initial concentration decreases exponentially with the distance from the point of release.

Partitioning

The deposition of particles only affects the sorbed fraction of the chemical, volatilization only affects the neutral dissolved fraction. Dissociated acids or bases usually neither sorb nor volatilize. Fractions are found as follows:

$$C_t = C_W + C_{Sorb} \qquad (6.7)$$

C_t : total concentration in water body (kg per m^3 water body)
C_W : concentration of dissolved chemical(kg per m^3 water body)
C_{Sorb} : concentration of chemical sorbed to particles (kg per m^3 water body).

The K_d value (Eq. 4.10) describes the concentration ratio between particle and water in local equilibrium:

$$K_d = C'_{Pa}/C'_W$$

C'_{Pa} : concentration sorbed to particles, unit kg substance/kg particles
C'_W : concentration dissolved, unit kg substance/kg water.
K_d : sorption coefficient, $K_d = K_{OC}OC$ (Eqs. 4.5 to 4.11b)

With P is the content of particles in water (g particles/m^3 or 10^{-6} t/t) it follows

$$C_{Sorb} = P \, 10^{-6} K_d C_W \qquad (6.8)$$

and thus $C_t = C_W + C_{Sorb} = C_W + P \, 10^{-6} K_d C_W$

Now the fraction of the chemical dissolved in water f_W is:

$$f_W = C_W/C_t = 1/(1 + P \, 10^{-6} K_d) \qquad (6.9)$$

The fraction of the chemical sorbed to particles f_{Pa} is

$$f_{Pa} = C_{Sorb}/C_t = P \, 10^{-6} K_d / (1 + P \, 10^{-6} K_d) \qquad (6.10)$$
$$f_W + f_{Pa} = 1, \text{ since } C_W + C_{Sorb} = C_t \quad (\text{Eq. } 6.7).$$

Processes are now calculated from C_t (total concentration) by using the factors f_W and f_{Pa}, which give the corresponding species.

Degradation

The main processes of degradation in water are hydrolysis, photolysis or biotic transformation. They are expressed as first order reactions or pseudo-first order reactions. WATER uses an aggregated degradation rate λ_{deg}, the sum of all rates. Sorption can influence the degradation. But values are not usually given for the sorbed or dissolved species; an overall rate for the river is normally taken. It is the user's responsibility to select an appropriate value for λ_{deg}. Note that

- hydrolysis is the sum of neutral-, base- or acid- catalyzed processes. When hydrolysis is of significance, the pH-value needs to be considered.
- aquatic photolysis is the sum of direct photolysis and indirect photochemical processes (usually more important). For procedures to estimate photolysis see Burns et al. (1982).
- biotic transformation is assumed to be of the first order. Other kinetics (higher order, Michaelis-Menten or Monod) need to be transformed into pseudo-first order.

Volatilization

The volatilization rate K_V can be estimated by using the 'two-film theory', which has already been mentioned in Chapter 4 (Whitman 1923, Mackay and Yeun 1980; Mackay and Paterson 1982; Burns et al. 1982; Trapp and Harland 1995).

For rivers it means: the water body is usually well mixed by turbulence (dispersion). The same applies to air. At the boundary between water and air, a stagnant zone or rather a laminar layer forms on both sides. These films can only be crossed by diffusion of the chemical. The process is thus described by the thickness of the boundary layers (diffusion path length), the concentration gradient between the media (the partition coefficient needs to be considered), and the diffusion velocities in both media. Both films' resistances are in series, and often the overall transfer is controlled by one layer.

The volatilization rate of very fugitive substances can be correlated to the reaeration rate (oxygen transfer). Reaeration rates have been examined intensively and are available for many hydrological situations.

In WATER, it is assumed that only the fraction dissolved in water can volatilize, and thus the elimination rate for the total concentration λ_V is:

$$\lambda_V = K_V f_W \tag{6.11}$$

λ_V : volatilization rate related to total concentration (time^{-1})
K_V : volatilization rate of the dissolved chemical (time^{-1})

Volatilization rate K_V (of the dissolved substance) is composed of the conductance (or resistance) of both films:

$$1/K_V = h[1/k_1 + 1/(K_{AW}k_g)] \tag{6.12}$$

K_V : volatilization rate of the dissolved chemical, time^{-1}
k_1 : conductance of the liquid film, length/time
k_g : conductance of the gaseous film, length/time
h : average river depth

In the WATER model, three different regression equations for the volatilization rate of oceans, lakes and rivers can be selected. For rivers (Southworth 1979):

$$k_1 = F\,0.2351u^{0.969}h^{-0.673}(32/M)^{1/2} \quad \text{(m/h)} \tag{6.13}$$

F = correction factor for high wind speeds:

$$F = \begin{cases} \exp[0.526(v - 1.9)] & \text{for } 1,9\,\text{m/s} < v < 5 \\ 1 & \text{for } v \leq 1.9\,\text{m/s} \end{cases} \tag{6.14}$$

$$k_g = 11.37(v + u)(18/M)^{1/2} \quad \text{(m/h)} \tag{6.15}$$

Units: Flow velocity u in m/s; depth h in m; v is the wind speed 10 cm above water level (m/s), M is the molar mass of the volatilizing substance (g/mol); relation of k_1 to oxygen ($M = 32$ g/mol), k_g to water ($M = 18$ g/mol).

v at a height of 10 cm may be calculated from v at 10 m from the logarithmic wind profile:

$$\frac{v(10\,\text{m})}{v(0.1\,\text{m})} = \frac{\log(10\,\text{m}/z_0)}{\log(0.1\,\text{m}/z_0)} \cong 2$$

z_0 is the roughness height, above rivers $\cong 0.001$ m (Burns et al. 1982).

The ratio k_g/k_1 is relatively constant, approximately 10^2 to 10^3 (Wolff and van der Heijde 1982), while the partition coefficient between atmosphere and water K_{AW} can vary by more than ten orders of magnitude. From Eq. 6.12 is can be seen that K_{AW} is the fate-determining parameter.

$K_{AW} < 4 \cdot 10^{-6}$: Volatility can be neglected, the substance volatilizes slower than water (K_{AW} of H_2O ca. $17 \cdot 10^{-6}$).

$4 \cdot 10^{-6} < K_{AW} < 4 \cdot 10^{-4}$: Volatilization from water is controlled by small diffusion through the gaseous layer.

$4 \cdot 10^{-4} < K_{AW} < 0.04$: Conductances of liquid and gaseous layer k_1 and k_g influence the results.

$0.04 < K_{AW}$: The substance is fugitive, volatilization is usually of great importance. K_V is only controlled by the liquid layer k_1. K_{AW} and k_g are of minor importance. K_V can be estimated from reaeration rates and molar mass.

Southworth's equation is only valid for medium or large rivers (depth about 3 m, flow velocity > 0.4 m/s). For lakes and deep, slow rivers or channels see Mackay and Yeun (1983), for the open sea refer to Liss and Slater (1974). All equations are also given in Trapp and Harland (1995), together with a field test.

Sedimentation

The exchange processes of a chemical between a water body and sediment have an extreme spatial and dynamic variability (see, e.g. Westrich 1988). In the WATER model, deposition, resuspension and diffusive exchange are not considered separately. If deposition is larger than resuspension, a net loss of the chemical for the fluid follows. Loss from the water body by net deposition is calculated from the growth of the sediment S (m/s or mm/a). Empirical values of S are:

$S < 0$	:	erosion, no net deposition; rivers with a high slope.
$S = 0$	:	sedimentation equilibrium; rivers are usually meandering.
$0 < S < 1$ mm/a	:	small deposition, slowly and normally flowing rivers.
$1\text{mm/a} < S < 3\text{mm/a}$	:	average deposition in lakes; very slow rivers, groynes, stagnant zones.
$S \gg 3\text{mm/a}$	:	close to emittents or sources of high particle input; before barrier weirs; eutrophic lakes (Dyck and Peschke 1983).

Sedimentation only affects the fraction of chemical sorbed to particles, therefore the factor f_{Pa} is required (Eq. 6.10).

$$\lambda_s = f_{Pa}\, c\, S\rho(1 - \varepsilon)/(h\, P\, 10^{-3}) \tag{6.16}$$

c	:	scaling factor, mm/a to m/d $= (10^{-3}/365)$
λ_S	:	net rate of loss to sediment (d^{-1})
S	:	sediment growth (mm/a)
h	:	water depth (m)
ρ	:	dry sediment density (kg/m^3)
ε	:	sediment porosity (vol/vol)
P	:	particle content in water (from Eq. 6.8: g/m^3, therefore factor 10^{-3})

Dynamic events such as deposition at low water or resuspension during a flood cannot be simulated using this approach.

Model equation

All elimination pathways are expressed as first order rate constants. Their addition gives the total elimination rate:

$$\lambda = \lambda_{deg} + \lambda_V + \lambda_S \tag{6.17}$$

(units of λ here d^{-1}). For concentration C (at x) it follows:

$$C(x) = C_0 e^{-\lambda x/u}$$

In steady-state, sources and sinks are equal. The total amount of substance m (kg) in the river segment is the integral of concentration C over length L, multiplied by the cross-sectional area A:

$$m = A \int_0^L C(x)dx \;=\; C_0 A u/\lambda \left[1 - \exp(-\lambda L/u)\right] \tag{6.18}$$

$$=\; C_0 Q/\lambda \left[1 - \exp(-\lambda L/u)\right]$$

The fluxes for the single processes follow from

$$dm/dt = -\lambda_i m \tag{6.19}$$

by taking λ_i of the process considered.

Bioconcentration in fish BCF

The steady-state concentration in biota (fish) close to the point of emission is estimated from the BCF (Eq. 4.13) and concentration C_0, with a correction for the (neutral) dissolved species:

$$C_{fish,max} = BCF\, C_0 f_W \tag{6.20}$$

At the end of the investigated stretch the (minimum) concentration is

$$C_{fish,min} = BCF\, C_{end} f_W \tag{6.21}$$

where C_{end} is concentration C at $x = L$. The density of fish is assumed to be identical to that of water (1 kg/L fresh weight). The uptake into fish is neglected for the mass balance.

Equilibrium concentration in sediment

Similarly, the equilibrium concentration in sediment is calculated from C_0 or C_{end} using the partition coefficient and factor f_W:

$$C_{S,max} \;=\; C_0 f_W K_{SW}$$

$$C_{S,min} \;=\; C_{end} f_W K_{SW}$$

where

$$K_{SW} = C_S/C_W = K_d \rho_S/\rho_W + \theta \qquad\qquad \text{(Eq. 4.8)}$$

Example 6.1: Volatilization of trichloroethene from the River Main

The River Main is a tributary of the Rhine. There are several large metal industry plants, situated mainly in Schweinfurt. Trichloroethene (Tri) has been used and emitted there (1990). What concentration of Tri can be found in the river water 300 km downstream, if 0.3 kg Tri is emitted daily, and no other significant inputs occur?

Average data of the River Main: flow velocity 0.5 m/s, depth 3 m, volume flow 105 m^3/s, width 70 m, average wind speed (at a height of 10 cm) 1 m/s, 50 g/m^3 particles with 10% OC.

Data of Tri (Tables 4.1 and 4.2): $K_{AW} = 0.5$, $K_{OW} = 195$, M = 131.39 g/mol, no significant degradation, no dissociation.

Initial concentration (transversely and vertically averaged):

$$
\begin{aligned}
C_0 = I/Q &= 0.3\,\text{kg/d}/(105 \cdot 86400\,\text{m}^3/d) = 33 \cdot 10^{-9}\,\text{kg/m}^3 \\
&= 33\,\text{ng/L}
\end{aligned}
$$

With the K_{OW} of 195, only a small sorption to particles and sedimentation is expected. We neglect both here ($K_d < 10$, $f_{Pa} < 1\%$).

The high partition coefficient between air and water indicates high volatility of the substance. $K_{AW} > 0.04$: The substance is very volatile, the volatilization is controlled by the liquid film conductance, k_1. The wind speed is below 1.9 m/s, no correction is necessary. The calculation simplifies to

$$
\begin{aligned}
k_1 &= 0.2351\,u^{0.969}\,h^{-0.673}\sqrt{(32/M)}\ \text{m/h} \\
&= 0.2351\ 0.5^{0.969}\ 3^{-0.673}\sqrt{(32/131.39)}\ \text{m/h} = 0.028\,\text{m/h} \\
1/K_V &= h[1/k_1 + 1/(K_{AW}\,k_g)]
\end{aligned}
$$

With $K_{AW}\,k_g \gg k_1$:
$K_V = k_1/h = 0.028\,\text{m}\,\text{h}^{-1}/3\,\text{m} = 0.0094\,\text{h}^{-1} = 0.23\,d^{-1}$

The half-life for volatilization is $\ln 2/K_V = 3$ days. Other loss processes are comparatively small and

$$
\begin{aligned}
C_{end} &= C_0\,e^{-\lambda t} \\
&= I/Q\,e^{-\lambda x/u}
\end{aligned}
$$

$x/u = 300\,000\,\text{m}/0.5\,\text{m/s} = 600\,000\,\text{s} = 6.9\,d$

$$
\begin{aligned}
C_{end} &= 33\,\text{ng/L}\,e^{-0.232\times 6.9} \\
&= 0.2\,C_0 = 6.6\,\text{ng/L}.
\end{aligned}
$$

An exponential decrease downstream is caused by volatilization, with a half-life of approximately 3 days, corresponding to a flow distance of 130 km. 300 km downstream, the volume flow has doubled and the concentration has additionally decreased by dilution. The concentration of Tri clearly does not reach the legal limit (for drinking water, FRG 1995) of 10 µg/L. This example is taken from Matthies et al. (1992), see also Trapp and Harland (1995).

Example 6.2: Chloroorganics in Finnish lakes

An example of the application of the WATER model is the investigation of the fate of chloroorganics emitted from a pulp mill in Finland (Trapp et al. 1994), Fig. 6.2, Table 6.1. Measured concentrations in the Äänekoski water course and in the waste water were available. The model was used to find elimination rates of chlorophenols and -guaiacols by inverse modeling (fitting of the elimination

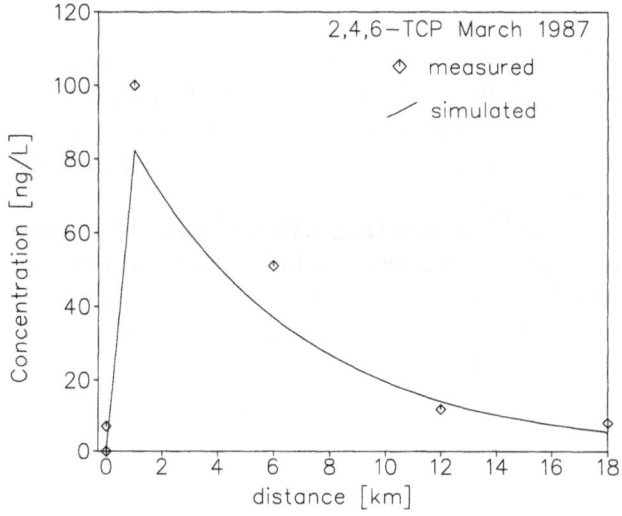

Fig. 6.2. Simulation of the concentration of 2,4,6- trichlorophenol, March 1987, Äänekoski; Trapp et al. (1994)

Table 6.1. Calculated mass balance of 2,4,6-trichlorophenol, March 1987, Äänekoski

Process		mass flux	[%]
Volatilization	:	0.0046 kg/d	1.7%
Sedimentation	:	3.75E-05 kg/d	0.014%
Degradation	:	0.25 kg/d	92.65%
Advection	:	0.015 kg/d	5.62%
Total	:	0.27 kg/d	100.00%

rates). Similar elimination rates of approximately 0.22 d^{-1} with a corresponding half-life of approximately three days were found for 2,4,6-tri-, 4,5,6-tri-, tetra-chloroguaiakol and for 2,4,6- trichlorophenol. Sedimentation and volatilization of these chemicals are small. A possible explanation for the similarities of rates is the elimination by biotransformation of the benthos bacteria; the rate is controlled by transport to the benthos.

In this example, annual average release of 2,4,6-TCP was taken as input. This explains the deviations between measured and simulated concentrations.

6.3
Dynamic Numerical TOXRIV Model

Purpose

The WATER model is usually not applicable for the simulation of chemicals' transport under real dynamic conditions. In most cases, neither steady-state nor constant hydrological conditions occur. In natural rivers, the flow velocity u and cross-section area A usually vary. Exchange processes with sediment and stagnant zones influence the transport.

An accident (pulse input) could be simulated with an analytical solution of the dispersion-advection equation. Numerical solutions are also adequate. Here we use the numerical solution scheme we have shown in Chapter 3, the explicit finite differences method with correction of numerical dispersion, as the basis of the dynamic numerical TOXRIV model.

Mass balance

Besides concentrations in the water body, TOXRIV also calculates dynamic concentrations in sediments and stagnant zones. The differential equations are coupled by exchange terms. Within one box, the chemical may undergo the same processes as in WATER, i.e. dilution, volatilization, degradation, loss to sediment and bioaccumulation. Additionally, some more processes are considered, namely, exchange with sediment including resuspension and exchange to stagnant water zones. The principal structure of a box in TOXRIV is shown in Fig. 6.3.

Analogous to the explicit finite differences method in Chapter 3, the river is divided into spatial segments $i = 1, \ldots, n$, and the time is divided into time intervals $j = 1, \ldots, m$.

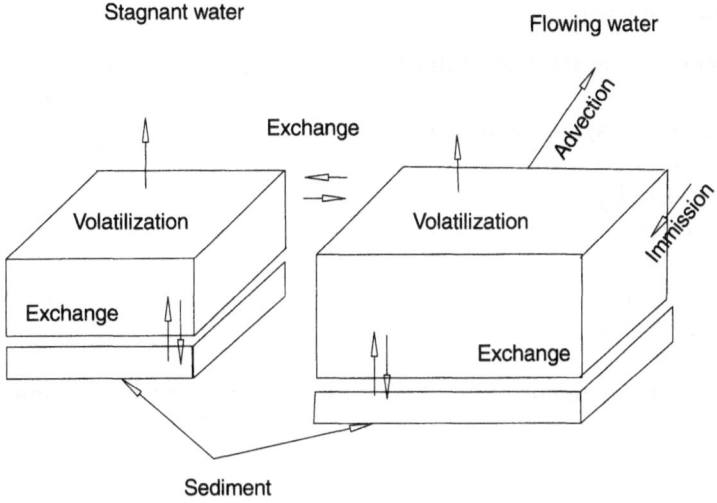

Fig. 6.3. Principal structure of a box in TOXRIV

The mass balance is calculated for every node i, j. With flowing water, stagnant water and sediment, this is:

Mass of substance in flowing water at place i and time interval $j + 1 =$
mass of substance at i and j
$+$ advection from the upstream segment $i - 1$ in time interval j with duration Δt
$-$ advection from the considered segment i in time interval j
$-$ volatilization to air in segment i in time interval j
$-$ loss by degradation in segment i in time interval j
$-$ loss to sediment of segment i in time interval j
$-$ loss to stagnant zone of segment i in time interval j
$+$ profits from sediment of segment i in time interval j
$+$ profits from stagnant zone of segment i in time interval j
$+$ input in segment i in time interval j

Longitudinal dispersion is not explicitly described. The numerical dispersion is set identical to the longitudinal dispersion by the selection of appropriate time intervals and segment lengthes.

The sediment is considered to be immobile; advective terms do not occur:

Mass of substance in sediment at place i in time period $j + 1 =$
mass of substance in sediment i, j
$-$ loss by degradation in i, j
$-$ loss to flowing water i, j
$+$ profits from flowing water i, j

Similarly for immobile stagnant water zones:

Mass of substance in the stagnant water zone at place i during time period $j + 1 =$
mass of substance in stagnant zone i, j
− loss by degradation i, j
− loss to flowing water i, j
− loss by volatilization
− loss to sediment
+ profits from flowing water i, j
+ profits from sediment

Mass of substance in sediment of the stagnant water zone at place i in time period $j + 1 =$
mass of substance in sediment i, j
− loss by degradation in i, j
− loss to stagnant water i, j
+ profits from stagnant water i, j

The mathematical description of the model, sensitivity analysis, data requirements and application are found in Matthies et al. (1992).

Input data required

The more complex the model, the more data it usually requires. TOXRIV requires 24 input data for each box, most of which vary with time and space. The data are not available in this resolution. The problem can be solved using the Monte Carlo analysis, which gives the sensitive parameters and the uncertainty of model results.

To show the influence of increasing model complexity on the calculated concentration of a pollutant, a (fictive) simulation of a pulse input of hexachlorobenzene into the Middle Rhine is shown in Fig. 6.4.

Hexachlorobenzene sorbs and volatiles. Volatilization, as well as sediment interactions, lead to a decrease in concentrations in flowing water. The effect of groynes (stagnant water zones in the Rhine) is an additional elimination, but mainly leads to a tailing of the peak.

A complex model such as TOXRIV, with its very high data requirement (and with it a high expenditure of time and money) should only be used when absolutely required by the problem.

In practice (e.g. alarm for drinking water supply following an accident), mainly the arrival and duration of the pollutant wave and approximate peak concentration are required. The model should be user-friendly, fast and not prone to errors. Simple analytical models, possibly interlinked with fate models for process estimation, are usually preferred (KHR 1991, Brüggemann et al. 1991).

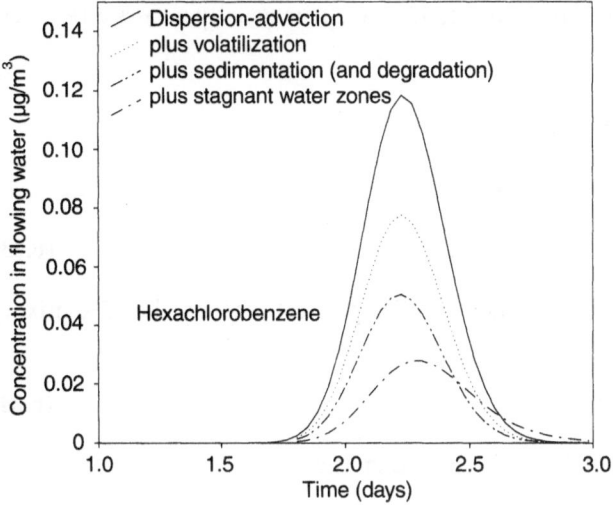

Fig. 6.4. Simulation of hexachlorobenzene in the Middle Rhine, successive addition of processes

6.4
Two-Dimensional Approach

Solution of the diffusion-advection equation for the y-direction

Until now, we have assumed immediate transversal mixing and only considered the flow direction. But in reality, it takes some time and distance to achieve mixing. The width of natural rivers is by far greater than the depth, and mixing in the y- direction takes much longer than in the z-direction. A particular problem of transversal mixing in rivers is the reflection of particles or molecules at the banks. The solution of the two-dimensional steady-state transport problem of a conservative substance emitted at one bank is (Benedict 1981):

$$C(x, y) = \frac{I}{hu \left[\pi D_y(x/u)\right]^{1/2}} \sum_{n=-\infty}^{\infty} \exp -\frac{(y - 2nB)^2}{4D_y(x/u)} \tag{6.22}$$

$C(x, y)$: vertically averaged steady-state concentration at x and y (kg/m³); h: average depth of the river (m); x: coordinate in flow direction (m); y: transversal coordinate (m), $0 < y < B$; B: average width of the river (m); I: input of substance (kg/s), here at $y = 0$; u: mean flow velocity (m/s), $u = Q/(hB)$; Q: volume flow (m³/s); D_y: transversal dispersion coefficient (m²/s), see below; n: number of reflections at the banks.

In principle, the equation should be summed up from $-\infty$ to $+\infty$, starting with 0, 1, -1, 2, -2.... For practical reasons, this is stopped when a new iteration contributes less than 1/1000 to $C(x, y)$. This usually occurs after a few steps.

Length of mixing distance

The mixing is complete when $C(x, y)$ is

$$C(x, y) = I/Q = I/(huB) = C_M \qquad (6.23)$$

Theoretically, this will never be the case. For pratical reasons, the 95% mixing distance is taken:

$$C(x, y)/C_M = 0.95 \qquad (6.24)$$

The corresponding length L (m) is (Benedict 1981):

$$L \cong 0.4 \, u \, B^2/D_y \qquad (6.25)$$

The most sensitive parameter is the width of the river.

Example 6.2: 95% mixing distance

Three cases: small river, medium river, large river

Case One: Small river in the lowlands (e.g. Hase in Osnabrück)
Width 5.0 m, depth 0.5 m, flow velocity 0.3 m/s, $D_y = 0.008 \, \text{m}^2/\text{s}$; 95% mixing after 370 m;
   In a small river, mixing occurs after a few hundred meters. It is rarely necessary to consider two dimensions.

Case Two: Medium-sized river
Width 100 m, depth 4 m, flow velocity 0.5 m/s, $D_y = 0.076 \, \text{m}^2/\text{s}$; 95% mixing after 26.5 km;
   In a medium-sized river (allowing for shipping, e.g. Main, Mosel, Weser), it takes more than 20 km for 95% mixing (Fig. 6.5).

Case Three: large river, stream (e.g. Rhine)
Width 400 m, depth 2.5 m, flow velocity 1 m/s, $D_y = 0.1 \, \text{m}^2/\text{s}$; 95% mixing after 633 km;
   With large rivers, such as Danube or the Rhine, it can take several hundreds of kilometers until transversal mixing is complete (Fig. 6.6). This, of course, has important consequences for water monitoring.

Trick: It is possible to calculate steady-state and one-dimensional $C(x)$ and consider the second dimension by multiplying $C(x)$ with the y-profile ($0 < y < B$). This gives $C(x, y)$.

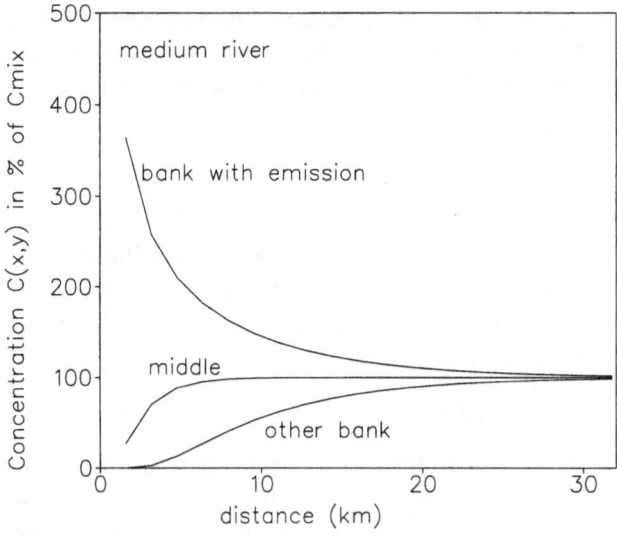

Fig. 6.5. Transversal mixing in a medium-sized river

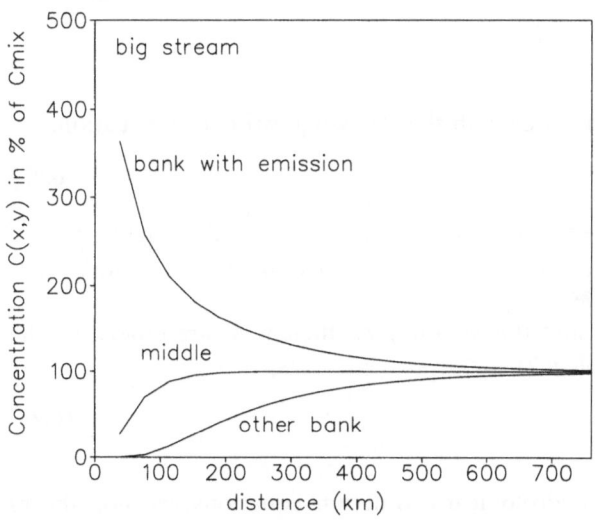

Fig. 6.6. Transversal Mixing in a big stream

6.5
How to Obtain Hydrological Data

Relations between the properties of a river

The hydrological properties of a river – water level, volume flow, flow velocity, width, depth, dispersion coefficient, but also reaeration rate, suspended particles, deposition and resuspension rate – are all related with one another. The use of master variables, such as the water level, allows the estimation of others – when the form of the relation is known. Empirical relations can be found by non-linear regression. Please refer to a good hydrology book.

Longitudinal dispersion

When only one dimension for the calculation of transport is used, the term D_x for dispersion in the x-direction is misleading, since an aggregated dispersion coefficient is used: The longitudinal dispersion coefficient D_L (m^2 s^{-1}). It can be estimated using the Fischer equation (Mazjik 1987):

$$D_L = \beta\, u\, B^2\, c\, h^{-1} g^{-1/2} \quad (\text{m}^2/\text{s}) \tag{6.26}$$

u: average flow velocity (m/s); B: river width (m); h: river depth (m); c: roughness coefficient of Brahms and de Chezy (m s$^{-1/2}$) = Chezy coefficient (a measure for roughness of the river bed, approximately $45\,\text{m s}^{-1/2}$ for the Rhine, Mazjik 1987); g: acceleration due to gravity (9.81 m^2/s); β: proportionality constant, $\beta = 0.002$ (Reichert and Wanner 1987) and $\beta = 0.001$ (own fit with official river data) for the Middle and Lower Rhine.

Transversal dispersion

Fischer (1979, p. 107) provided us with the following estimation equation:

$$D_y = a\, h\, u_* \tag{6.27}$$

u_*: shear velocity at the river bed, $u_* = u\sqrt{g/c}$; g: acceleration due to gravity (9.81 m^2/s); c: Chezy coeffizient (ms$^{-1/2}$); a: proportionality constant, Fischer (1979, p. 112): $a = 0.6 \pm 0.3$;

The Chezy coefficient c and the Manning coefficient M are related by the hydraulic radius R (see Table 6.2).

$$c = M\, R^{1/6} \tag{6.28}$$

$R = hB/(2h + B)$.

Note that the units of hydrological estimation equations are not always consistent.

Table 6.2. Manning – coefficients (m$^{1/3}$/s) (Hermann 1984)

lowland rivers, no obstacles	60
with stones and plants, meandering	40
Lower Rhine	40 to 45
mountain rivers with flint stones	45
plus rocks	36

6.6
Contaminants in Waste Water Treatment Plants

In principle, water in a waste water treatment plant (WWTP) can also be considered as flowing surface water. An emitted pollutant undergoes the same processes as in a river, i.e. advection, sorption, volatilization and, of course, degradation. When steady-state and equilibrium between phases are assumed, then a simple model for the behavior of organic chemicals in a WWTP is:

Change of mass in the WWTP =
+ inflow
− degradation
− sedimentation to the sewage sludge and removal
− volatilization
− outflow

Mathematical description (see below for assumptions):
inflow $= Q_{in}C_{in}$
degradation $= \lambda V C_t$
sedimentation $= Q_S C_S = R Q_{in} K_d C_W$
volatilization $= Q_A C_A = Q_A C_W K_{AW}$
outflow $= Q_{out} C_W = Q_{in} C_W$

Assumptions:
Immediate equilibrium between water, sludge and air blown into the system; degradation in water and sludge is the same. The amount of sludge is constant (sludge forming is removed at the same rate); outflow is only water; homogeneous mixing; steady-state.

Parameters and variables:
Q_{in} = inflow of waste water (m^3/s); C_{in} = concentration in inflow (kg m^{-3}); λ = first order degradation rate (s^{-1}); C_W = concentration in water; C_t = total concentration in water and sludge (kg m^{-3}); Q_S = flux of sludge (growth) (m^3/s); R = ratio between sludge and water (t of sludge per m^3 water); K_d = concentration ratio of the chemical between sludge and water (m^3/ton = cm^3/g); Q_A = air volume blown in (m^3/s, depending on the type of WWTP, it can be calculated from the change in biological oxygen demand); C_A = equilibrium concentration in air (calculate from K_{AW}); outflow $Q_{out} = Q_{in}$; V = volume of the WWTP (m^3).

The total amount of substance m_t in the WWTP is

$$m_t = V C_t = V_W C_W + V_S C_S = V_W C_W + R V_W K_d C_W \qquad (6.29)$$

The mass balance equation of m_t is

$$dm_t/dt = Q_{in}C_{in} - \lambda V C_W (1 + RK_d)$$
$$-R Q_{in} K_d C_W - Q_A C_W K_{AW} - Q_{in} C_W \qquad (6.30)$$

Steady-state $dm_t/dt = 0$ and solving for C_W gives

$$C_W = \frac{Q_{in}C_{in}}{\lambda V(1 + RK_d) + RQ_{in}K_d + Q_A K_{AW} + Q_{in}} \tag{6.31}$$

Admittedly, this is an extremely simplified model of a WWTP. It can, however, facilitate the deduction of significant processes (compare Exercise 6.2).

Exercises on Chapter Six

6.1 Estimate the volatilization rate of

a) Tetrachloroethene (C_2Cl_4), $M = 165.8$, $\log K_{OW} = 2.87$, $K_{AW} = 0.83$
b) Atrazine ($C_8H_{14}ClN_5$), $M = 215.7$, $\log K_{OW} = 2.64$, $K_{AW} = 8 \cdot 10^{-9}$

River data: Flow velocity 0.5 m/s, depth 2.5 m, volume flow 100 m³/s (width 80 m), average wind speed at a height of 10 cm is 1.8 m/s, 50 g/m³ particles with 10% OC.

6.2 Sewage purification plant
A biological sewage purification plant has the volume 10 m × 10 m × 2 m. The waste water flow is 0.01 m³/s. In the sewage purification plant, 10 kg (= 10 liters) of sludge per m³ water are formed. It settles and is removed. The organic carbon content is 10%. The aeration is 0.1 m³ air/s.

What is the fate of pollutants in this WWTP?

a) Make a model for the mass balance of a pollutant in the WWTP. Calculate steady-state concentrations of the pollutants in the outflow for:
b) the plasticizer DEHP (Di-ethyl-hexyl-phthalate)
   ($\log K_{OW} = 5.0$; $K_{AW} = 0.3 \cdot 10^{-3}$; degradation rate $= 1\,d^{-1}$, concentration in inflow $= 1$ mg/L);
c) 1,4-dichlorobenzene (a compound used in toilet stones)
   ($\log K_{OW} = 3.4$; $K_{AW} = 0.1$; degradation rate $= 0.1\,d^{-1}$, concentration in inflow $= 1\,\mu$g/L).
d) Is it reasonable to measure the cleaning power of the WWTP by comparing in- and outflow concentrations?

6.3 Call up CemoS, load trichloroethene, load the WATER model and calculate the mass balance with the standard scenario. Do the same with atrazine. Discuss the results.

Chapter 7:
Transport and Transformation of Compounds in Soil

7.1
Soil and Its Functions

The status of soil as an environmental medium worth protecting was recognized later than the other two relatively uniform media, air and water. In contrast to air and water, soil constitutes a heterogeneous assembly with animate cavities, the formation, development and preservation of which is dependent on the contribution of organisms or biological processes. Soil protection comprises most importantly the conservation of its functions. This requires on the one hand, the provision of the preservation of soil as a resource, and on the other hand, measures to clean up existing damage. Soil should be conceived as three-dimensional sectors of ecosystems. It represents the area in which the lithosphere, the hydrosphere, the atmosphere and the biosphere under the formation of a fifth sphere, the 'pedosphere', penetrate one another. If this definition is taken as a basis, soil would be limited to the animated sectors of the earth's surface that are not saturated by water. Soil constitutes structural and functional elements of terrestrial ecosystems. Soil therefore comprises reaction vessels in which the processes of substance metabolism and energy consumption are connected to each other. All functions of soil can be derived from this definition, as well as the justification of its worthiness (Table 7.1).

When dealing with pollutants in soil the production and regulation functions are most important. What is meant here by production function is the capacity of soil to supply primary producers (green plants) with water and nutrients and to serve as a root area for plants, also and especially under the aspect of cultivation with the aim of producing utilizable biomasses. Pollutant transport and behavior is mostly affected by the regulation function of soil. This includes the transport, filtering, buffering, transformation and accumulation of water, heat and substances. Soil conveys energy and substance exchanges using transport processes between groundwater, the atmosphere and neighboring ecosystems. Soil is a life zone for a multitude of various plants, animals and microbial organisms. The regulation function of soil is considerably based

Table 7.1. Soil and its functions

1	Ecological Functions
1.1	Agricultural and forestry production
1.2	Filter, buffer and transformation function
1.2.1	Physical buffering ("Storage capacity for water and heat")
1.2.2	Physical-chemical filtering and buffering ("Sorption of substances")
1.2.3	Biological-biochemical transformation
1.3	Gene protection and gene reserve function
2	Techno-industrial, socio-economic and cultural functions
2.1	Infrastructure ("Provision of ground and space")
2.2	Raw materials ("Ores, building materials, etc.")
2.3	Cultural functions ("Archaeology, palaeontology")

on the substance metabolic rate and growth capacity of these organisms. In their totality, soil organisms are the carriers of the degradation, conversion and construction of substances in soil. Due to a high diversity of species, their interactions contribute to the stability of ecosystems in which, amongst other things, they degrade toxic substances.

Soil protection has a two-fold meaning. First, the input of contaminants should be as low as possible so as not to endanger soil life and their ecological functions. Second, soil protection also means preventing or reducing the dangers that could lead to the output of contaminants which could affect neighboring ecosystems, such as ponds or other biotopes. Simultaneously, a good regulation function of soil means that no or very small amounts of substances should enter the ground water, thus protecting drinking water reserves. The take-up of harmful substances in plants is the first step towards contaminating food and fodder. Gaseous emissions from soil can be transported by the wind and either influence the atmosphere as a trace gas or reach other terrestrial or acquatic ecosystems by deposition.

Sources, Inputs and Occurence of Soil Contamination

Figure 7.1 shows the sources and most important input pathways of harmful substances to soil. Four major sources can contribute to the contamination of soil:

1) Atmospheric emissions from traffic, industry or combustion can be deposited onto the soil surface as particles or gaseous compounds or onto the vegetation canopy and then via litter fall to the ground. This has led to the considerable contamination of large areas with a multitude of various types of pollutants, i.e. heavy metals, persistent organic chemicals, acids

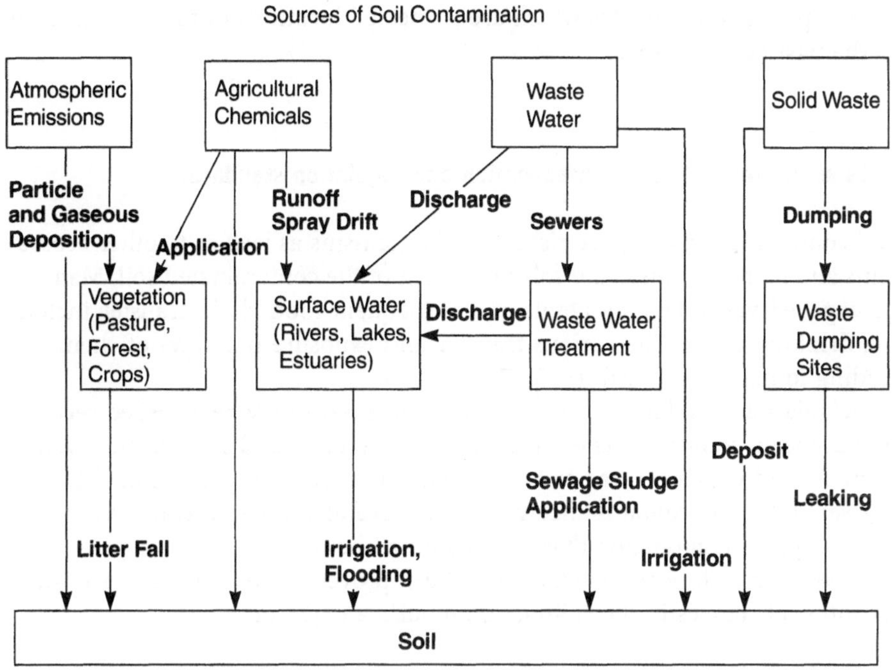

Sources of Soil Contamination

Fig. 7.1. Sources of soil contamination

and bases. Airborne pollutants can be transported over large distances thus causing contamination in remote areas.

2) Agricultural chemicals, i.e. pesticides and mineral fertilizers as well as manure are applied in almost all agricultural systems. In Germany, the percentage of drinking water with a concentration above 20 mg/l nitrate increased from 1.9% in 1915 to 32.6% in 1989 (UBA, 1995). The natural level of nitrate in water is less than 10 mg/l.

3) Waste water is either discharged directly to fresh water (rivers, lakes) or salt water (estuaries) or after purification in a sewage treatment plant. If polluted water is used for irrigation or if natural or artificial flooding occurs, chemical pollutants can enter the soil. Another important pathway which has led to considerable amounts of heavy metals and persistent xenobiotics in soil is the application of sewage sludge to soil as a fertilizer (Drescher-Kaden et al., 1990).

4) Solid waste from industrial production as well as from households has been dumped into shallow soils or deposited onto the ground. When we refer to toxic waste from the past, we mean 'toxic waste depositions', i.e. abandoned and disused disposal sites, rubbish dumps, etc. as well as 'toxic waste locations', i.e. the sites of disused plants. In Germany, about 140 000 waste dumping sites were known in 1993. In addition, it can be assumed that there

are approx. 100 000 sites which are suspected locations of toxic waste from the past (UBA, 1995).

Risks associated with soil contamination and regulation standards

Soil contamination can pose risks to soil organisms as well as to other ecosystems and man by transfering substances out of the contaminated soil. Man can be exposed via uptake by plants and subsequent food chain transfer including drinking water. The various transfer and exposure pathways of chemicals leading to man are shown in Fig. 7.2.

Regulation standards for soil contaminants have only been issued recently in European countries. The Dutch list of reference and intervention values (now A- and C-values) was one of the first attempts to regulate soil pollution (Visser, 1993). It contains mostly metals, several inorganic compounds and a few organic compounds that are graded as being dangerous or potentially dangerous to soil, or those which have been proven to be important persistent harmful substances in the soil/plant/animal/man system.

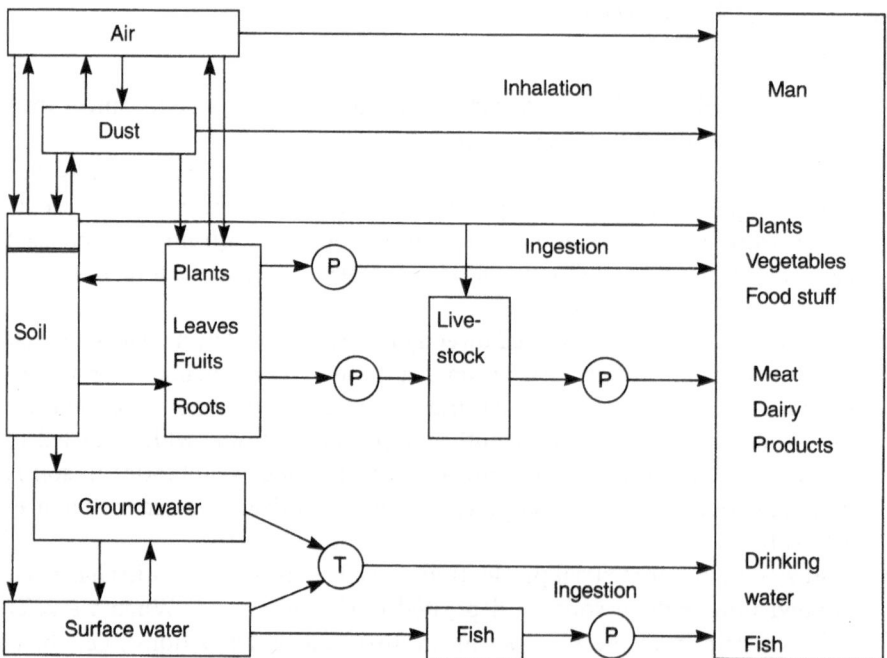

Fig. 7.2. Exposure Pathways for Health and Environmental Risk Assessment; **P** = Processing, **T** = Treatment

A draft version of a federal soil protection act is intended to be passed in Germany during the present legislative period, i.e. by 1998. The act would regulate both soil contamination and toxic waste from the past. Soil pollution is evaluated by taking into consideration the various exposure and utilization pathways. A distinction is made between test and intervention values. If a test value is exceeded, a site investigation has to be carried out to ascertain whether a harmful change to the soil has occurred. Intervention values involve measures which need to be taken to dispose of, secure, or clean-up the contamination in question.

Several regulations on drinking water, sewage sludge application, etc. are used to assess the risks associated with soil contamination. The EU Technical Guidance Document for risk assessment of new and existing chemicals considers soil contamination as a source for the exposure of man via food chain and terrestrial organisms (EU 1996; van Leeuwen and Hermens 1995).

7.2
Transport Processes of Substances in Soil

Properties of soil

'Soil' is the top layer of the earth. It is used by vegetation as an anchor, mineral source and water reservoir. The rooted layer is only a few meters deep. This thin layer is the basis of nearly all terrestrial life!

Soil is composed of minerals, gas, water, biota and humic substances. Soil types are differentiated according to the particle size. Sandy soils (sand 63 to $2\,000\,\mu m$ diameter) have a high fraction of coarse pores. They are well aerated, but only have a small storage capacity for water and a high water conductance. The fraction of organic substances (humic) is usually small. The mobility of xenobiotics in sandy soils is comparatively high. Clay soils (clay $< 2\,\mu m$ diameter) have the finest particles. They have the highest total porosity of all soils, but rarely have coarse pores. Aeration is not good and the conductance is small. But when clay soils become dry, they crack. Preferential flow occurs along these cracks and is responsible for water and substance transport into deeper layers. Loamy soils are usually most fertile. They have the highest storage capacity for water and are also well aerated.

The transport and fate of substances in soil are controlled by

- advection (with soil water) and leaching into ground water
- dispersion/diffusion in gas and water filled pores
- sorption to the soil matrix
- biotic and abiotic degradation or transformation
- uptake by plants
- volatilization into the atmosphere.

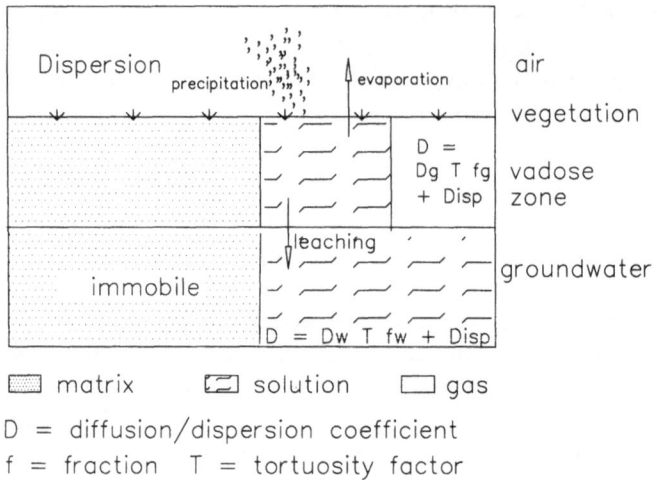

Fig. 7.3. Phases and transport processes in the groundwater/soil/vegetation/air system

Mixing of chemicals may also occur by soil organisms, e.g. mice, rabbits, and earth worms. Liquids such as oil may form their own phase. The latter processes are not considered here. Soils are nearly always covered by vegetation, and roots take up water and chemicals (see Chapter Nine).

The parametrization of these processes in soil differs considerably from that of flowing waters: the flow velocity is orders of magnitude smaller. Sorption to the soil matrix plays a major role. Time scales are far greater (years). This implies that metabolic degradation and transformation play a significant role.

Advection

In temperate climates, vertical water flow in soil occurs after precipitation. The flow direction and velocity are determined by the water potential. About 2/3 of the precipitated water is taken up by plants and is transpired (Central Europe). The following steady- state calculation of the water balance is simple, but rather fictive: We assume an annual precipitation of 767 mm (1 mm = 1 liter/m^2), 3/4 volatilizes and 1/10 is runoff:

+ precipitation (2.1 mm/d)
− evaporation (1.6 mm/d)
− runoff (0.2 mm/d)
− storage (steady-state, = 0 mm/d)
= infiltrating water (0.3 mm/d)

The specific flow q (filter velocity) to the ground water is 0.3 mm/d.

Dispersion and diffusion

The flow velocity through porous media varies according to inhomogenities of the pores. This leads to hydrodynamic dispersion. The dispersion coefficient D_{disp} (m^2/s) depends on the distance passed and is calculated from the dispersion length L_{disp} (m) multiplied by the filter velocity q (Bear 1972):

$$D_{disp} = L_{disp} q \qquad (7.1)$$

L_{disp} is an empirical parameter, varying with soil type and increasing with distance. For flow velocities $u = q/\theta \leq 1$ cm/d, values of L_{disp} between 0.5 to 5 cm (flow distance < 10 m) were found. A default value of 5 cm is used in the SOIL model. This is an average value of sandy and loamy soils (Matthies et al. 1987). θ is the fraction of water-filled pores (volumetric water content).

The effective molecular diffusion coefficient in the aqueous soil solution $D_{W,eff}$ (m^2/s) is only relevant with small filter velocities ($u = q/\theta \leq 10$ cm/d). It is calculated from the diffusion coefficient D_W of the substance in water, reduced by a labyrinth or tortuosity factor T (Jury et al. 1983):

$$D_{W,eff} = T\, D_W \qquad (7.2)$$

T is the tortuosity factor (Millington and Quirk 1961):

$$T = \theta^{10/3}/\varepsilon^2 \qquad (7.3)$$

ε is the total porosity of the soil (vol/vol).

Both D_{disp} and $D_{W,eff}$ are added to give the apparent dispersion/diffusion coefficient $D_{W,a}$:

$$D_{W,a} = D_{disp} + D_{W,eff} \qquad (7.4)$$

Analogous to the aqueous phase, the effective diffusion coefficient in the gas phase $D_{G,eff}$ (m^2/s) is calculated:

$$D_{G,eff} = D_G(\varepsilon - \theta)^{10/3}/\varepsilon^2 \qquad (7.5)$$

For estimates of D_W and D_G, see Chapter Eleven.

Partitioning

Only the dissolved fraction (f_W) of the chemical is transported by advection of the soil water and dispersion in the aqueous phase. Diffusion in the soil gas phase only affects the gaseous fraction (f_g). The fraction sorbed to the matrix (f_m) is immobile.

We calculate the fractions by assuming instantaneous local equilibrium. The total concentration (per m^3 bulk soil) is:

$$C_t = C_M + C_W + C_G$$

C_t, C_M, C_W and C_G: concentration of a chemical per m^3 bulk soil; t = total, M = sorbed to the matrix, W = in soil water and G = gaseous.

Now we relate the concentrations to a cubic meter of the corresponding phase, expressed as X:

$$C_t = X_M + \theta X_W + (\varepsilon - \theta) X_G$$

X is the mass of chemical per m^3 phase.

Introducing the partition coefficients gives:

$$X_M/X_W = K_{MW} = K_d \rho_B/\rho_W = OC\, K_{OC} \rho_B/\rho_W \qquad \text{(Eq. 4.6 = 7.6)}$$

$$X_G/X_W = H/(RT) = K_{AW} \qquad \text{(Eq. 4.3 = 7.7)}$$

X_M is the concentration of the chemical sorbed to the soil matrix (kg/m^3 dry soil), X_W is the concentration in soil water (kg/m^3 soil water) and X_G is the concentration of the gaseous chemical in the soil gas phase (kg/m^3 soil gas phase). K_{MW} is the dimensionless linear partition coefficient between soil matrix and water, ρ_B is the bulk density of the dry soil (kg/m^3). Now, the total concentration in soil C_t (kg/m^3 bulk soil) may be expressed in the following form:

$$C_t = K_{MW} X_W + \theta X_W + (\varepsilon - \theta) K_{AW} X_W \qquad (7.8)$$

and with $C_W = \theta X_W$

$$f_w = C_W/C_t = \theta/[K_{MW} + \theta + (\varepsilon - \theta) K_{AW}] \qquad (7.9)$$

and

$$f_g = C_G/C_t = (\varepsilon - \theta) K_{AW}/[K_{MW} + \theta + (\varepsilon - \theta) K_{AW}] \qquad (7.10)$$

$$f_m = C_M/C_t = K_{MW}/[K_{MW} + \theta + (\varepsilon - \theta) K_{AW}] \qquad (7.11)$$

The fractions describe the amount of dissolved, gaseous or sorbed chemical per cubic meter bulk soil:

$$C_W = f_w C_t$$

$$C_G = f_g C_t$$

$$C_M = f_m C_t$$

$$f_w + f_m + f_g = 1$$

The concentration in the corresponding phase *per m^3 phase*, X, can be calculated from the volume ratio between the corresponding phase and the total

Table 7.2. Experimental and calculated transport velocities relative to water; Tinsley (1979)

Name	K_d	f_w	experimental
2,4-D	0.32	0.66	0.69
Simazine	1.35	0.32	0.45
Atrazine	1.72	0.27	0.47
Diuron	4.85	0.11	0.24
Chloroxuron	50	0.012	0.09
Paraquat	200	0.003	0.00
DDT	2 430	0.0003	0.00

volume. For the matrix (dry soil), C_M is equal to X_M. The concentration in soil water $X_W = C_W/\theta$.

The movement of a chemical relative to water is described by the 'retention factor (R_f)' (Tinsley 1979), or by the 'retardation factor (R_d)' (Bear 1979):

$$R_f = u_c/u_w = 1/R_d$$

u_c : advective transport velocity of a chemical in soil
u_w : flow velocity of water in soil, $= q/\theta$

The advective transport velocity of a chemical in soil can also be found from the fraction dissolved in soil water:

$$u_c = u_w f_w$$

In our nomenclature, R_f is identical to f_w. The retardation factor R_d is inverse to it. A comparison to experimental values is given in Table 7.2.

Bulk soil density

The density of the dry bulk soil ρ_B (kg/m^3) is related to the total porosity ε and the content of organic matter OM (Benzler et al. 1982):

$$\rho_B = (1 - \varepsilon)[OM \cdot 1400 + (1 - OM) \cdot 2650] \tag{7.12}$$

Organic matter and organic carbon content OC are related by (Benzler et al. 1982):

$$OM = 1.724 \cdot OC \tag{7.13}$$

Wet weight – dry weight

In laboratory, soil mass is usually measured as dry weight. The most usual unit of concentration is mg/kg. If not stated otherwise, it should be dry weight. In environmental models, however, the concentration is usually based on volume.

The SI-unit of concentration is kg/m^3. Errors may occur during transfer from one unit to another.

Consider a soil with a dry bulk soil density of ρ_B is 1300 kg/m^3. The volumetric water content θ be 0.3 m^3 water per m^3 soil, resp. 300 kg water per m^3 of soil. Thus, the wet soil density is 1600 kg/m^3.

Now if you have 1 kg substance per m^3 soil, it depends on the water content (density given for the dry soil) how much you have in 1 kg of soil:

dry soil: 1 kg / 1300 kg = 0.000769 kg/kg = 769 mg/kg (dry wt.)
wet soil: 1 kg / 1600 kg = 0.000625 kg/kg = 625 mg/kg (wet wt.)

This is somewhat confusing, since for both wet and dry soil, the substance concentration is 1 kg per m^3, because the volume of the soil changes normally only negligibly with the water content. The mass of the 0.3 m^3 air, that is replaced by the water, is about 0.390 kg and is negligible, too.

The other case:
You have a given concentration of 1 mg/kg (dry wt.).
This corresponds to 1.3 g/m^3 soil (dry). This is identical to 1.3 g/m^3 soil (wet).
If the concentration is 1 mg/kg (wet wt.), this corresponds to 1.6 g/m^3 soil.

We leave the trivial question to the reader, how concentrations in the phases of the soil (dry matrix, solution, gas phase; either per cubic meter or per kg) are calculated.

Diffusion/advection equation for transport in soil

The concentration in soil resulting from vertical transport is calculated using the one-dimensional diffusion/dispersion-advection equation, combined with a first-order sink term:

$$\frac{\partial C_t}{\partial t} = -u\frac{\partial C_t}{\partial z} + D\frac{\partial^2 C_t}{\partial z^2} - \lambda C_t \tag{7.14}$$

The transport terms can be written separately for the soil water and soil gas:

$$\frac{\partial C_t}{\partial t} = -u_w\frac{\partial C_W}{\partial z} + D_{W,a}\frac{\partial^2 C_W}{\partial z^2} + D_{G,eff}\frac{\partial^2 C_G}{\partial z^2} - \lambda C_t$$

$$= -u_w\frac{\partial(f_w C_t)}{\partial z} + D_{W,a}\frac{\partial^2(f_w C_t)}{\partial z^2} + D_{G,eff}\frac{\partial^2(f_g C_t)}{\partial z^2} - \lambda C_t$$

and for constant fractions f_w, f_g

$$= -u_w f_w\frac{\partial C_t}{\partial z} + (D_{W,a} f_w + D_{G,eff} f_g)\frac{\partial^2 C_t}{\partial z^2} - \lambda C_t$$

$$= -u_c\frac{\partial C_t}{\partial z} + D_c\frac{\partial^2 C_t}{\partial z^2} - \lambda C_t$$

Solutions of the diffusion/advection equation from other problems (e.g. rivers, groundwater or of nonsorbing tracers in soil) can be used by replacing u by $u_w \cdot f_w$ for advection and D by $(D_{W,a} f_w + D_{G,eff} f_g)$ for diffusion plus dispersion.

7.3
SOIL – Analytical Solutions for Vertical Transport in Soil

Analytical solutions of the partial differential equation (7.14) depend on initial and boundary conditions. Here are a few examples, for further solutions see Genuchten and Alves (1982).

Pulse input

For the following conditions

- diffusion/dispersion coefficient is constant
- input m_0 (kg) is at $z = 0$ and $t = 0$ (pulse input)
- area A (m^2) is constant,
- θ, the fraction of water-filled pores, and total porosity ε are constant
- all soil water is mobile
- flow velocity u (m/s) is constant
- the soil is infinite, $C(\infty, t) = 0$

it follows for the concentration in depth z at time t (see Eq. 3.31):

$$C_t(z, t) = \frac{m_0/A}{\sqrt{4\pi Dt}} \exp\left[-\frac{(z - ut)^2}{4Dt}\right] \exp(-\lambda t) \tag{7.15}$$

$C_t(z, t)$: total concentration at z and t (kg/m^3)
z : vertical coordinate in flow direction (m)
t : time after release (s)
m_0 : initial amount (kg)
A : surface area (m^2)
D : diff./dispersion coefficient (m^2/s) = $D_{W,a} f_w + D_{G,eff} f_g$
u : flow velocity in z-direction (m/s) = $u_w f_w$
λ : first-order reaction rate constant (s^{-1})

The concentration in soil water (kg/m^3 water) is $X_W = C_t f_w/\theta$.

The solution does not consider reflection at the soil surface. It is not valid when most of the substance is close to the surface.

Contaminated layer

When a contaminated layer of soil is the initial condition, the solution is found by integrating many point sources (Crank, 1979, p. 14). The initial concentration of the chemical in the layer $-h < z < h$ is $C = C_0$, anywhere else $C = 0$; no degradation; other conditions as above. The solution is:

$$C(z, t) = \frac{C_0}{2} \left\{ \mathrm{erf} \frac{z - ut + h}{\sqrt{4Dt}} - \mathrm{erf} \frac{z - ut - h}{\sqrt{4Dt}} \right\} \tag{7.16}$$

With a (constant) degradation: replace C_0 by $C_0 \exp(-\lambda t)$.

The solution requires a 'standard function' of mathematics, the so-called *error function* $\mathrm{erf}(z)$. It is found in tables and can be approximated. It is defined as:

$$\mathrm{erf}(z) = \frac{2}{\sqrt{\pi}} \int_0^z \exp(-\eta^2) d\eta$$

The function has the properties

$$\mathrm{erf}(-z) = -\mathrm{erf}(z); 1 - \mathrm{erf}(z) = \mathrm{erfc}(z); \mathrm{erf}(0) = 0; \mathrm{erf}(\infty) = 1;$$

$\mathrm{erf}(z)$:   *error function*
$\mathrm{erfc}(z)$:   *complementary error function*

An approximation of $\mathrm{erf}(z)$ is given in Abramowitz and Stegun (1972) for $0 \le z \le \infty$

$$\mathrm{erf}(z) = 1 - (a_1 b + a_2 b^2 + a_3 b^3) \exp(-z^2) + \varepsilon(z)$$

$$|\varepsilon(z)| = \text{absolute error} \le 2.5 \cdot 10^{-5};$$

$$b = 1/(1 + pz); \quad p = 0.47047;$$

$$a_1 = 0.3480242; \quad a_2 = -0.0958798; \quad a_3 = 0.7478556$$

Continuous injection

Continuous injection into the top of a soil column has the initial conditions

$C = 0$ for $t \le 0$ and $z \ge 0$
$C = C_0$ for $t > 0$ and $z = 0$

The lower boundary condition is again $C(\infty, t) = 0$ for all t.

The solution is (Bear 1972, cited in Bear 1979, p. 268):

$$C(z, t) = \frac{C_0}{2} \exp[uz/(2D)] \left\{ \exp(-z\beta) \mathrm{erfc} \frac{z - t\sqrt{u^2 + 4\lambda D}}{\sqrt{4Dt}} \right.$$

$$\left. + \exp(z\beta) \mathrm{erfc} \frac{z + t\sqrt{u^2 + 4\lambda D}}{\sqrt{4Dt}} \right\} \tag{7.17}$$

$$\beta = \sqrt{u^2/(4D^2) + \lambda/D}$$

For $t \to \infty$ the steady-state solution is ($\mathrm{erfc}(\infty) = 0$, $\mathrm{erfc}(-\infty) = 2$):

$$C(z) = C_0 \exp[uz/(2D) - z\beta]$$

Analytical solution for volatilization from soil

For substances with $K_{AW} \gg 10^{-4}$, diffusion in gas pores is the fastest transport process (when air-filled pores are present, of course). For those substances, the vapor pressure is usually high, and volatilization from soil becomes significant. Air and soil are two spaces with different uptake capacity and diffusion coefficients.

Soil $(z < 0)$
C = total concentration in bulk soil = C_B
D = effective gaseous diffusion coefficient of the gaseous fraction in soil = $D_{G,eff} f_g = D_B$
initial condition: semi-infinite contaminated soil, $C_B(t = 0, z < 0)$ = constant = $C_B(0)$

Air $(z > 0)$
C = concentration in air = C_A
D = diffusion coefficient in gases: $D = D_G$ or D_A
initially clean air
$C_A(t = 0, z > 0)$ = constant = 0

A molecule that reaches the boundary may diffuse either into air or back into soil. At the boundary $(z = 0$, soil surface), the concentration is in local equilibrium, described by the partition coefficient K:

$$K = K_{AB} = C_A(z = 0)/C_B(z = 0)$$

At the boundary, the flux out of the soil is equal to the flux into the air (and vice versa):

$$D_B \partial C_B / \partial z = D_A \partial C_A / \partial z \quad (z = 0)$$

For this problem, Jost (1960, p. 69) found as early as 1937 a standard solution (Jost 1960, p. 69, see also Crank 1979, p. 39):

$$C_B(z, t) = \frac{C_B(0)}{1 + K\sqrt{D_A/D_B}} \left\{ 1 + K\sqrt{(D_A/D_B)} \operatorname{erf} \frac{-z}{2\sqrt{D_B t}} \right\} \qquad (7.18a)$$

$$C_A(z, t) = \frac{K\,C_B(0)}{1 + K\sqrt{D_A/D_B}} \operatorname{erfc} \frac{z}{2\sqrt{D_A t}} \qquad (7.18b)$$

At the boundary $(z = 0)$, the concentration is constant, because $\operatorname{erf}(0) = 0$ and $\operatorname{erfc}(0) = 1$

$$C_B(0, t > 0) = \frac{C_B(0)}{1 + K\sqrt{D_A/D_B}} \qquad (7.19a)$$

$$C_A(0, t > 0) = \frac{KC_B(0)}{1 + K\sqrt{D_A/D_B}} \qquad (7.19b)$$

For volatilization from soil into vegetation see Trapp and Matthies (1997).

Example 7.1: Transport of an ideal tracer

Let the steady-state water balance of a soil be

+ precipitation 2.1 mm/d
− evaporation 1.6 mm/d
− runoff 0.2 mm/d
± stored water 0 mm/d
= specific flow 0.3 mm/d = filter velocity q

With 30% water-filled pores, the flow velocity $u = q/0.3 = 1$ mm/d. The dispersion coefficient D_{disp} is $L_{disp}q = 0.05$ m $\times 0.3 \cdot 10^{-3}$ m/d $= 1.5 \cdot 10^{-5}$ m^2/d. The diffusion coefficient for a molar mass M of 128 g/mol is (Chapter Eleven, Eq. 11.13) $D_W = 8.65 \times 10^{-5}$ m^2/d. The effective diffusion coefficient with $\varepsilon = 0.5$ is

$$D_{W,eff} = D_W \theta^{10/3}/\varepsilon^2 = 6.25 \cdot 10^{-6} \text{m}^2/\text{d}$$

The apparent diffusion/dispersion coefficient $D_{W,a}$ is the sum of both:

$$D_{W,a} = D_{disp} + D_{W,eff} = 2.125 \cdot 10^{-5} \text{m}^2/\text{d} = \text{D}$$

For an ideal tracer, no sorption and reaction ($f_w = 1$, $\lambda = 0$) occur. If it is continuously injected, Eq. 7.17 can be applied. The concentration is shown in Fig. 7.4. The unit of concentration depends on $C(z = 0, t)$, here we might take concentration in soil solution (mg/L). Even an ideal tracer needs a long time to leach down, in this case about 5 years to 2 m depth. Sorption ($f_w < 1$) would decrease the mobility of the chemical. Therefore, chemicals in soil with a half-life of below one year are usually degraded before they reach ground water, except under unfavorable conditions. The comparison of Figs. 7.4 and 3.2 (movement in a river) makes the differences in times scales obvious (years as opposed to days).

7.4
Comment

The approaches presented for the transport of chemicals in soil in Sects. 7.1 to 7.3 assume a steady-state water balance. The results of simulations with the diffusion-advection equation indicate that sorbing substances do move slowly (Fig. 7.4). In reality, precipitation is discontinuous, as is infiltration. The water movement also depends on the storage capacity of the soil. In particular after heavy rain, water transport in soils may be rather rapid. Therefore, within short time periods much more substance is leached than predicted with the steady-state water balance. The transport in macro pores (e.g. in cracks, along roots, in the tunnels of mice and rabbits) may play a considerable role for

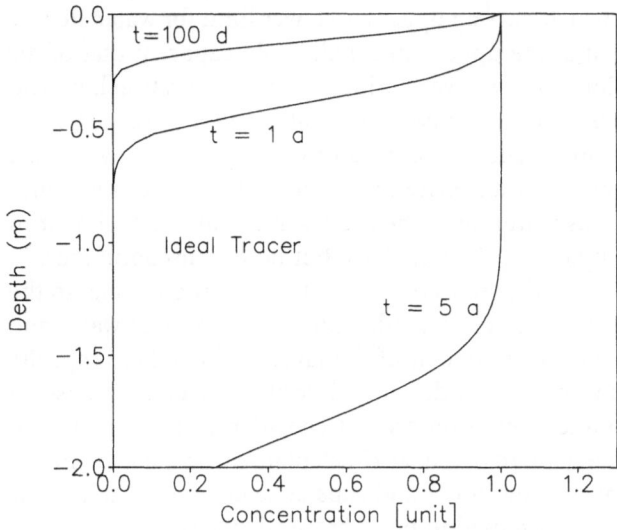

Fig. 7.4. Movement of an ideal tracer in soil; example calculated with CemoS (version 1.0)

substance movement. Indeed, chemicals can be found in deeper soils due to such a hydraulic 'short circuit', and may well enter groundwater (Feher et al. 1991, Persicani 1993). Pesticides and other compounds whose continuous transport is low, actually can be found in groundwater (e.g. atrazine). In the next chapter, a discrete cascade model is shown which simulates unsteady water and substance transport by infiltration and capillary rise.

In practice, numerical solutions are often used. They allow more variable boundary conditions (Behrendt et al., 1990, Huston and Wagenet, 1992). Nonetheless, analytical solutions will always be important for testing the numerics and, of course, for didactical reasons. Additionally, the influence of parameters and processes (sensitivity) can be seen directly. No numerical problems will occur. But sometimes, the approximations of $erf(z)$, $erfc(z)$ and even the exponential function are defective, depending on the approximation method, compiler and computer.

Exact calculations of groundwater contamination are very uncertain for larger environmental segments due to the heterogenity of soils and the transient conditions.

7.5
Discrete Cascade Model BUCKETS

In the models described above, transport in soil was calculated by use of the dispersion-advection equation. A different approach is the discrete cascade model or the tipping bucket approach. The soil is divided into a series of tipping buckets or cascades, which have an upper and lower limit for water storage capacity. The water content at the upper limit is the field capacity, that of the lower limit the wilting point (if there is vegetation) or the evaporation limit (no vegetation). No continuous flow is assumed. The soil layers are considered as buckets that can be filled up to field capacity, after which they tip. By putting these layers in series, tipping buckets arise that transport water and solvents. The model is discrete and discontinuous. The approach is somewhat similar to the cascade model in example 2.3 (Chapter Two), but flow is discontinuous. In the continuous cascade model, the transport at any time is proportional to the mass of the substance. In the discrete cascade model, transport of water and solute only occurs when the water content of the layer is above field capacity (Richter 1990). The tipping bucket model only describes advective transport with soil water. Diffusion and dispersion are not considered. There is no distinction between pore sizes; therefore it also calculates transport in macro pores. The model does not consider water and substance uptake by plants. The degradation of compounds is assumed to be of first-order kinetics.

The reverse process, vertical transport upwards to the soil surface by capillary transport and evaporation, can also be simulated. Water evaporates from the soil surface until the lower limit of water content (wilting point) is reached. From then on, water plus solutes are also drawn upwards by capillary forces.

The principle (for positive water balance) is shown in Fig. 7.5.

The BUCKETS model in CemoS is based on the tipping buckets approach.

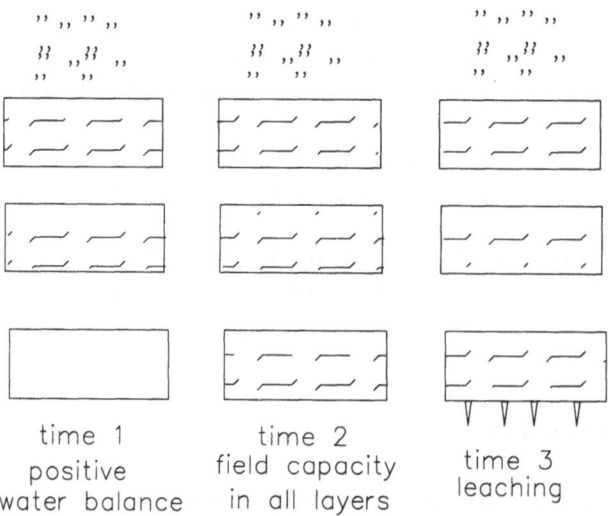

time 1
positive
water balance

time 2
field capacity
in all layers

time 3
leaching

Fig. 7.5. Principle of the tipping buckets approach

Infiltration and leaching

The soil column is divided into n layers of equal thickness d (m), and the simulation time is divided into m time steps of the same duration Δt (d). At the beginning of the simulation, all layers have the initial volumetric water content θ_0 (m^3/m^3) and therefore the initial amount of water W_0:

$$W_0(i) = \theta_0(i)d \tag{7.20}$$

where W_0 = (initial) amount or volume of water per thickness of layer and area m^2 (unit W = m^3 water / m^2 soil = m); i indicates the soil layer (counting downwards).

Precipitation falls on the top soil layer. The water partly runs off and evaporates, the rest infiltrates as long as the water balance is positive. The infiltrated amount of water per time interval (filter velocity q) is calculated from the water balance:

$$q = \text{precipitation} - \text{evaporation} - \text{runoff}$$

In the tipping buckets model, a dynamic water balance is used. For every time interval, q can be recalculated. As usual in meteorology, all parameters are normalized to an area of one square meter. This gives the unit L/(m^2 d) = 10^{-3} m/d, when the unit of precipitation is liter per day.

Water movement is simulated in discrete time intervals of duration Δt. The flux of the infiltrated water through all layers is calculated per time interval. After the first time interval Δt, the new amount of water in the top soil layer $W(1)$ is the sum of the initial amount of water $W_0(1)$ plus the infiltrated amount of water:

$$W(1) = W_0(1) + q \cdot 10^{-3}\Delta t \tag{7.21}$$

Now the question follows whether the bucket is full, i.e. the amount of water is above field capacity:

$$W(1) \geq FC\,d\,? \tag{7.22}$$

or

$$W(1) - FC\,d = \delta \tag{7.23}$$

In BUCKETS, the simplification is made that the field capacities of all layers are equal. If the inequation is not fulfilled, or when $\delta < 0$, no water flux occurs to the second layer, and the amount of water in the lower layers remains constant for this time interval. Then the new water content of the top soil layer $\theta(1)$ is

$$\theta(1) = W(1)/d \tag{7.24}$$

Otherwise, the absolute of the difference $|\delta|$ is added to the next lower layer. The new amount of water in the first and second layer is:

$$W(1) = FC\,d \tag{7.25a}$$

and

$$W(2) = W_0(2) + [W(1) - FC\,d] \tag{7.25b}$$

The calculation is repeated for the second layer

$$W(2) \geq FC\,d\,? \tag{7.26}$$

and for the third, ... and so on. If the inequation is not fulfilled for one of the layers, the water contents of the lower layers remain constant. Otherwise, the surplus of water is transported downwards. The iteration scheme for the flux of water from layer i to the lower layer $i+1$ is

$$W(i) - FC\,d = \delta \tag{7.27}$$

If $\delta < 0$, then

$$W(i) = W_0(i) + [W(i-1) - FC\,d] \tag{7.28a}$$

and for all layers $\geq i$:

$$W(i+1) = W_0(i+1). \tag{7.28b}$$

If $\delta \geq 0$, then

$$\begin{aligned} W(i) &= FC\,d \tag{7.29}\\ W(i+1) &= W_0(i+1) + [W(i) - FC\,d] \tag{7.30} \end{aligned}$$

The new volumetric water content of all soil layers $i = 1, \ldots, n$ is

$$\theta(i) = W(i)/d \tag{7.31}$$

When water is transported in this way to the lowest soil layer n, the surplus of water after one time step, $W_{out,1}$, can leach out of the soil column. For one time interval, $W_{out,1}$ is:

$$W_{out.1} = W(n) - FC\,d \tag{7.32}$$

The first time interval is now over, and the next one begins with Eq. 7.21. The complete scheme to Eq. 7.32 is calculated again. The iteration is repeated during the whole simulation time. After m time steps in total, the leached amount of water is:

$$W_{out,m} = \sum_{j=1}^{m} W_{out,j} \tag{7.33}$$

Evaporation and capillary transport

If the water balance is negative, the water moves upwards. This is the case when evaporation leads to a smaller water content in the top layer than in the lower layers. Water evaporates from the soil surface until the lower limit of water content (wilting point) is reached. The wilting point is defined as the volumetric water content that is the transpiration limit of plants. Below this limit, the water potential of the soil matrix is below that of the roots. For soils not covered by vegetation, the evaporation limit EL (m^3 water per m^3 soil) is the water content where capillary movement stops. In BUCKETS, the evaporation limit EL is assumed to be constant for all layers. For the top soil layer it holds

$$W(1) = W_0(1) - |q| \cdot 10^{-3} \Delta t \tag{7.34a}$$

Now we ask whether EL has already been reached:

$$W(1) \geq EL\, d \quad \text{or} \quad W(1) < EL\, d\,? \tag{7.34b}$$

If $W(1) > EL\, d$, for all $i > 1\, W(i) = W_0(i)$. In the case of $W(1) = EL\, d$, $W(i) = W_0(i)$ for all i. In the case of $W(1) < EL\, d$:

$$W(2) = [W_0(2) - EL\, d] - |W(1) - EL\, d| \tag{7.35a}$$

If $W(2) < EL\, d$, then $W(2) = EL\, d$, and the next layer is asked and so on:

$$W(i) = [W_0(i) - EL\, d] - |W(i - 1) - EL\, d| \tag{7.35b}$$

As long as $W(i) < EL\, d$, $W(i)$ is set to $W(i) = EL\, d$, and the procedure is continued for the next layer $i + 1$, until a layer k is found, $k > i$, where $W(k) \geq EL\, d$. For this layer it holds

$$W(k) = EL\, d \quad \text{for} \quad W(k) = EL\, d \tag{7.36a}$$

$$W(k) = W_0(k) - |W(k - 1) - EL\, d| \quad \text{for} \quad W(k) > EL\, d \tag{7.36b}$$

Water thus moves upwards until the amount of water given by the water balance is evaporated. Analogously to infiltration, all time steps m have iteratively been gone through and the total amount of evaporated water is summed up. When there is insufficient water in the soil column, the procedure is stopped when EL is reached in the lowest layer. The new volumetric water content and the evaporated amount of water is calculated as in the previous section.

Substance transport

Dissolved compounds are transported advectively, as described in Sect. 7.2. In the tipping buckets approach, neither dispersion and diffusion nor transport in soil air is considered. The approach should not be used for compounds with high volatility ($K_{AW} > 10^{-3}$), but it is applicable for ions, e.g. nitrates, most pesticides and others.

Substances can enter the soil surface through precipitation and infiltrate. Only the dissolved fraction of the chemical is mobile. The partition coefficient between the soil matrix and water is given in Eq. 7.6. Since the gas phase fraction can be neglected for chemicals with $K_{AW} \ll 10^{-3}$, Eq. 7.8 is simplified, and the fraction of chemical $f_w(i)$ dissolved in the soil layer i is

$$f_w(i) = \theta(i)/[K_{MW} + \theta(i)] \tag{7.37}$$

Substance and water movement occur simultaneously at every time interval. The initial concentration in the soil layers i is $C_0(i)$ (bulk soil concentration).

In our nomenclatura, the concentration in soil solute (kg substance per m^3 solute) is found from:

$$C(\text{solute}) = C(\text{bulk soil}) \cdot f_w/\theta$$

Only the solved chemical can be transported with leachate or evaporating water (compare Sect. 7.2 Partitioning).

In the first time interval, the concentration in infiltrating water C_I (kg/m^3 water) leads to an additional amount of chemical in the top layer. When vertical water flux occurs, substance is lost from the top layer. Additionally, substance may be lost by degradation. Thus, the equation for the new concentration in the top layer $C(1)$ is:

$$
\begin{aligned}
C(1) \;=\;& C_0(1)\exp(-\lambda\Delta t) \\
+\;& C_I q \cdot 10^{-3}\Delta t/d \\
-\;& C_0(1)f_w(1)/\theta(1) \cdot [W(1) - FC\,d]/d
\end{aligned} \tag{7.38}
$$

For the lower soil layers $i = 2,\ldots,n$:

$$
\begin{aligned}
C(i) \;=\;& C_0(i)\exp(-\lambda\Delta t) \\
+\;& C_0(i-1)f_w(i-1)/\theta(i-1) \cdot [W(i-1) - FC\,d]/d \\
-\;& C_0(i)f_w(i)/\theta(i) \cdot [W(i) - FC\,d]/d
\end{aligned} \tag{7.39}
$$

For the calculation of $W(i)$ for positive water balance (infiltration) see Eqs. 7.27 to 7.30.

Capillary increase during negative water balance can also lead to an upward substance flux:

$$
\begin{aligned}
C(i) \;=\;& C_0(i)\exp(-\lambda\Delta t) \\
+\;& C_0(i+1)f_w(i+1)/\theta(i+1) \cdot [W(i+1) - EL\,d]/d \\
-\;& C_0(i)f_w(i)/\theta(i) \cdot [W(i) - EL\,d]/d
\end{aligned} \tag{7.40}
$$

Water evaporates at the top layer, but the substance remains in the soil, because only compounds with a negligible potential for volatility are considered.

For the next time period, the new $C_0(i)$ is calculated from the momentary $C(i)$.

If a water surplus leaches out of the lowest soil layer, it may carry the chemical with it. The leached amount of compound $S_{out,1}$ after time step $j = 1$ is

$$S_{out,1} = W_{out,1} C_0(n) f_w(n)/\theta(n) \qquad (7.41)$$

The total amount of substance leached out from the soil column in the whole simulation time $S_{out,m}$ is the sum of the leached amounts in time intervals $j = 1, \ldots, m$:

$$S_{out,m} = \sum_{j=1}^{m} S_{out,j} \qquad (7.42)$$

Stability of the solution

The numerical scheme is identical to the explicit finite difference method (Sect. 3.8). The 'flow velocity of a substance particle' u for constant $W(i)$, $f_w(i)$, $\theta(i)$ is

$$u = [W(i) - FC\,d] f_w(i)/\theta(i) \quad \text{(infiltration)} \qquad (7.43)$$
$$u = [W(i) - EL\,d] f_w(i)/\theta(i) \quad \text{(evaporation)} \qquad (7.44)$$

The numerical dispersion is

$$D_n = 1/2[u\,d - u^2\Delta t] \qquad (7.45)$$

The Courant-number CR (must be ≤ 1) is $u\Delta t/\Delta x$. BUCKETS gives a warning and stops the simulation when CR is above 1.

By appropriate selection of $CR = 1$ or $\Delta t = d/u$, no numerical dispersion occurs. But since $W(i)$, $f_w(i)$ and $\theta(i)$ vary for each layer, numerical dispersion cannot be avoided. BUCKETS calculates the maximal numerical dispersion and gives it as an output value. It should be below or in the same order of amount as the diffusion coefficient of the substance (see SOIL).

Comparison to the SOIL model

Compared to the analytical solutions of the dispersion-advection equation that are implemented in the SOIL model, the advantage of BUCKETS is that it allows a dynamic water balance and variable input conditions, and that transport beyond the soil surface is impossible. The main disadvantage is that transport by diffusion and dispersion is neglected. With BUCKETS, the initial conditions of the solutions in Sect. 7.3 can also be simulated. When only the top soil layer is contaminated and all other layers are initailly clean, the conditions correspond to a pulse input function. If initial concentrations in some of the layers are identical, the solution is similar to a contaminated layer. A permanent substance input with the infiltrating water without precontamination of the soil corresponds to a continuous injection. With BUCKETS, combinations of these three cases among others, and insteady water balances may be simulated.

Exercises on chapter seven

7.1 Which chemical diffuses faster in soil?

The soil consists of 30% water-filled pores θ, 20% air filled pores ε, 2% organic carbon OC. q (filter velocity) 1 mm/d, dispersion length 5 cm, soil density $\rho = 1.3 \, g/cm^3$, pH $= 7$.

a) Benzene (C_6H_6), $M = 78.12$; $\log K_{OW} = 2.1$; $K_{AW} = 0.23$
b) Phenol (C_6H_6O), $M = 94.11$; $\log K_{OW} = 1.48$; $K_{AW} = 2.2 \cdot 10^{-5}$, $pKa = 9.9$
   (acidic).

7.2 Load all substances from the CemoS standard database, calculate the chemicals' advective velocity u_c and rank them.

Chapter 8:
Atmospheric Transport Models

8.1
Introduction

Chemical substances are released into the atmosphere from several sources: traffic, industry, households, heating systems, power stations, paintings, pesticide applications ...

In the atmosphere, chemicals are transported by diffusion/dispersion and advection and are eliminated by deposition or photodegradation.

8.2
Model Approaches AIR and PLUME

Two approaches are presented: the calculation of atmospheric transport and fate following area and point-source emission. Atmospheric and meteorological conditions are taken as constant (Trenkle and Münzer, 1987). Both models are part of SAMS (Screening Assessment Model System, Matthies et al. 1992a). For the area source, a steady-state box model with dilution by wind and elimination by deposition and degradation is used. For point emissions, a local plume model is used, including dilution by atmospheric dispersion (Gaussian type). The potential human dose is estimated from average inhalation.

Box model for area sources

Airborne pollutants are often emitted from (quasi-)area sources. Typical examples are diffuse urban emissions or the spraying of pesticides. In the one-compartment box model, the release area is rectangular and is assumed to be the sum of many sources (Fig. 8.1). The atmospheric mixing height z_0 multiplied by the release area $x_0 \cdot y_0$ is the target compartment. The resulting steady-state concentration in air C_0 is expressed as:

$$C_0 = I/(u\, y_0\, z_0 + v_{dep}\, x_0\, y_0 + \lambda V) \tag{8.1}$$

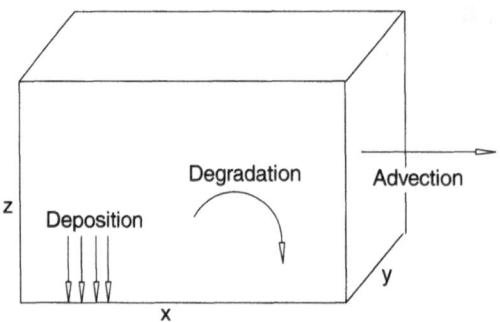

Fig. 8.1. Schematic representation of the box model AIR

Concentration = input/(advection + deposition + degradation)

C_0 :   steady-state concentration (kg/m^3)
I :   continuous release flow (kg/s)
x_0 :   length of box in wind direction (m)
y_0 :   width of box laterally to the wind direction (m)
z_0 :   atmospheric mixing height (m)
u :   vertically averaged wind speed (m/s)
v_{dep} :   overall deposition velocity (m/s), Table 8.1
λ :   total photolysis rate (s^{-1})
V :   volume of the box ($x_0\ y_0\ z_0$) (m^3)

Plume model for point sources

Concentrations downwards from point sources are calculated using a Gaussian model. It is a solution of the three-dimensional dispersion-advection equation (Fig. 8.2). Advection is due to movement with wind, dispersion is caused by atmospheric turbulence ('eddy diffusion').

When dispersion in x-(wind)direction and elimination are neglected, the concentration at distance x from the continuous point source is:

$$C(x, y, z, H) = \frac{I}{2\pi\sigma_y\sigma_z u} \exp \frac{-y^2}{2\sigma_y^2} \tag{8.2}$$

$$\cdot \left\{ \exp \frac{-(z - H)^2}{2\sigma_z^2} + \exp \frac{-(z + H)^2}{2\sigma_z^2} \right\}$$

The equation is used in the German *TA-Luft* (1986), a technical guidance which regulates emissions into air.

σ_y, σ_z: standard deviations of the Gaussian distribution function for directions y and z, parametrized with x (m)

H: effective release height (m) (depending on x, see TA-Luft).

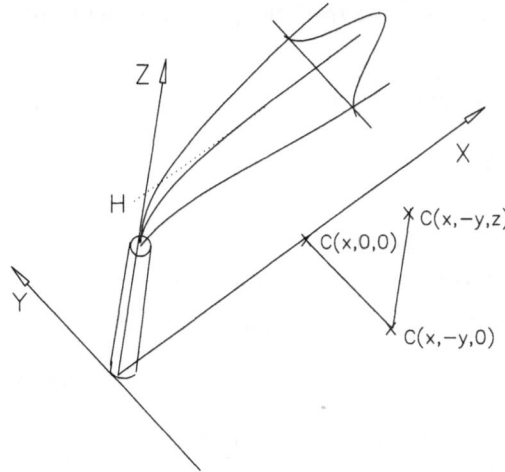

Fig. 8.2. Coordinate system of the Gaussian model

Some simplifications can be made. Concentration at the soil surface ($z = 0$) is:

$$C(x, y, 0, H) = \frac{I}{\pi \sigma_y \sigma_z u} \exp \left[\frac{-y^2}{2\sigma_y^2} - \frac{H^2}{2\sigma_z^2} \right] \tag{8.3}$$

Concentrations at the soil surface along the centre of the plume (maximal surface concentration, $y = z = 0$) are:

$$C(x, 0, 0, H) = \frac{I}{\pi \sigma_y \sigma_z u} \exp \frac{-H^2}{2\sigma_z^2} \tag{8.4}$$

The standard deviations of the Gaussian plume are the lateral and vertical 'dispersion coefficients' σ_y and σ_z. They are expressed as functions of x:

$$\sigma_y = a x^p \quad \text{and} \quad \sigma_z = b x^q \tag{8.5ab}$$

The coefficients were empirically determined and are given in the TA-Luft (1986). Long-term frequencies of wind speed, precipitation and Pasquill's stability classes were measured in Karlsruhe (Vogt, 1980). The neutral stability class dominates in Germany (62%). The parameters for neutral stability are used as default values in CemoS: atmospheric mixing height $z_0 = 500$ m, wind speed $u = 5$ m/s, coefficients $a = 0.640$; $p = 0.784$; $b = 0.215$; $q = 0.885$. Precipitation is 2.1 mm/d.

Gas-particulate partitioning

Chemicals can deposit from the plume to the surface. The deposition pathway of organic compounds is controlled by the partitioning between the gas and particle phases. The adsorbed fraction f_p follows from the partition coefficient between gas and particle K_{gp}

$$f_p = c_{par}/(c_{gas} + c_{par}) = 1/(1 + K_{gp}) \tag{8.6}$$
$$K_{gp} = c_{gas}/c_{par}$$

where

c_{gas} is the concentration of the gaseous chemical (kg/m^3 air) and
c_{par} is the concentration of the adsorbed chemical (kg/m^3 air).

Junge (1975) presented a model for adsorption to aerosols:

$$K_{gp} = p_s/(c\,A_P) \tag{8.7}$$

p_s : saturation vapor pressure of the liquid compound (Pa)
A_P : total surface area of aerosols (cm^2/cm^3 air)
c : parameter (Pa cm) = 17 Pa cm

Parameter c is not constant, but depends on aerosol and compound properties. The surface area of aerosols varies between $4.2 \cdot 10^{-7}$ for clean continental background air and up to $1.1 \cdot 10^{-5}$ cm^2/cm^3 for urban areas. A typical background value in the USA is $1.5 \cdot 10^{-6}$ cm^2/cm^3 (Bidleman, 1988). When semi-volatile organic chemicals with a solid state at given temperature (e.g. PCB and PCDD/F) are investigated, the vapor pressure of the subcooled liquid should be taken instead of that of the solid compound (Mackay et al. 1986). It follows from the Junge-equation that compounds with a vapor pressure of about 10^{-4} Pa at room temperature have similar gaseous and sorbed fractions. Temperature has a strong influence. A typical example is 2,3,7,8-TCDD ('Seveso dioxin') with a vapor pressure of the subcooled liquid of $9.4 \cdot 10^{-5}$ Pa at 25°C (Rordorf, 1992); under background conditions, between 25% and 75% are sorbed to particles (McLachlan, 1992).

Dry deposition

Gas and dry particle deposition are calculated with deposition velocities:

$$v_d = (1 - f_p)v_{d,gas} + f_p v_{d,par} \tag{8.8}$$

v_d : total dry deposition velocity (m/s)
$v_{d,gas}$: gaseous dry deposition velocity (m/s)
$v_{d,par}$: dry deposition velocity of particles (m/s)

The dry deposition velocities of organic chemicals have typical values in the range of a few mm/s and below (Bidleman, 1988). It is assumed that diffusion through a laminar boundary layer is the controlling process. This allows the estimation of deposition velocities from reference values v_d:

$$v_{d,gas} = v_{d,gas}(ref)\sqrt{\frac{M(ref)}{M}} \tag{8.9}$$

M is the molar mass (g/mol). A value of 5 mm/s for the reference compound, $M(ref) = 300$ g/mol, is used in CemoS (from Thompson, 1983).

Aerosols in ambient air are classified into three modes according to their size:

nucleation mode ($< 0.1\,\mu$m)
accumulation mode (0.1 to 1 μm)
sedimentation mode ($> 1\,\mu$m).

Depending on aerosol size, the residence time in air differs considerably. Small particles of the nucleation mode coagulate rather quickly, forming particles with diameters between 0.1 and 1.0 μm. These have the longest residence time in the atmospheric mixing layer (> 1 week), they accumulate and may be transported some thousand kilometers.

Larger particles of the sedimentation mode are quickly sedimented, they have a residence time of approximately 10^3 s and fall after a few kilometers, since gravity is higher than atmospheric turbulence ('Stokes Law'). In CemoS-AIR, the default deposition velocity $v_{d,par}$ of aerosols is 10 mm/s.

Wet deposition

The wet deposition of particle-bound and gaseous substances is proportional to the precipitation intensity. The scavenging of gases is determined by the partition coefficient air to water K_{AW} (Whelpdale, 1982). The wet deposition velocity of particle bound substances is calculated with an empirical washout ratio W_p:

$$v_w = \frac{(1 - f_p)P/K_{AW} + f_p W_p P}{8.64 \cdot 10^7} \tag{8.10}$$

v_w	:	wet deposition velocity (m/s)
P	:	precipitation intensity (mm/d)
K_{AW}	:	partition coefficient air to water K_{AW} (−)
W_p	:	washout ratio of particles (kg/m^3 rain : kg/m^3 air) $= 2 \cdot 10^5$
$1/(8.64 \cdot 10^7)$	:	scaling factor, mm/d to m/s

With an average precipitation rate of 2.1 mm/d, wet deposition velocity of particles is approximately 5 mm/s. This is in the same order of magnitude as dry deposition velocity.

Photolysis

Photolysis is the main process for the degradation of pollutants in air. New problems may occur from photoreaction ('photosmog'). Three mechanisms are of great significance:

– degradation by OH radicals (most important)
– degradation by ozone
– direct photolysis by sunlight.

In Eq. 8.1 (box-model), the sum of all three processes gives the first-order rate constant of degradation. For most of the organic chemicals, the reaction with OH radicals is the fastest:

$$R - X + OH \cdot \rightarrow XOH + R$$

This is a second-order reaction. The pseudo-first order rate is

$$\lambda = \lambda'_{OH}[OH\cdot] \tag{8.11}$$

λ'_{OH} :   second-order rate (cm^3/s/radicals)
$[OH\cdot]$:   average concentration of OH radicals in the mixing layer (radicals/cm^3)

An average default value of the radicals' concentration during the day is $[OH\cdot] = 10^6$ radicals/cm^3, at night 0 radicals (radicals themselves are also formed by photon collision), on average $= 5 \cdot 10^5$ radicals/cm^3.

Total deposition to the ground

The total flux by deposition d in kg/(m^{2}s) to the ground is calculated from the steady-state concentration and the deposition velocities:

$$d = v_{dep}C_0 = (v_d + v_w)C_0 \tag{8.12}$$

Table 8.1 gives ranges for deposition velocities of particles with different diameters.

Table 8.1. Deposition velocities of particles, TA-Luft (1986)

class	diameter μm	v_{dep} mm/s
$i = 1$	< 5	1
$i = 2$	5 to 10	10
$i = 3$	10 to 50	50
$i = 4$	> 50	100
	unknown	70

Potential inhalation dose

The potential inhalation dose for persons living downwind from the stack can be calculated from the steady-state concentration at the ground together with the inhalation per time period:

$$D = iC_0 \tag{8.13}$$

D : annual potential dose (kg/a)
i : annual inhalation (m^3/a)

A conservative estimate of inhalation i is 8 760 m^3/a, or 1 m^3/h.

Exercises on chapter eight

Benzene is a major air pollutant. The average emission in Germany is 187 kg per square kilometer and year (see chapter Five).

8.1 Calculate with the box model AIR

Concentration in the box
Dose in the box (kg/a)
Deposition in the box kg/(m^2 a)

Data benzene: vapor pressure $p = 10\,100$ Pa; molar mass $M = 78.11$ g/mol; $K_{AW} = 0.23$; input 0.5 kg/d; degradation rate 0.0475 d^{-1}.
Box: height $z_0 = 500$ m, length $= x_0$ and width $y_0 = 1000$ m. No advection, wind speed none (0 m/s), precipitation $= 2.1$ mm/d, particle surface area $= 1.5 \cdot 10^{-6}$ cm^2/cm^3, particles' dry deposition velocity 10 mm/s, dry deposition of a reference gas (M is 300 g/mol) $= 5$ mm/s, inhalation $= 20$ m^3/d (You can use the MIF-File EX81.MIF).

8.2 In chapter 5, we calculated 'Benzene in Germany' with a Level 2 model. Now, call CemoS Level 2, and enter those values given above (be careful with throughfluxes; note that Input here is kg/a; you can use the MIF-File EX82.MIF).

8.3 The results of exercise 8.1 and 8.2 differ significantly. Why is the concentration in the AIR box-model so much lower? Why is the concentration calculated with Level 2 higher than the one calculated in Chapter Five?

8.4 You had been working very late. At night, you bought goods at a gasoline station. The air at the station contained benzene (50 μg/m^3). You buy butter (250 g, volume 0.3 Liter). Benzene is a cancerogen.

a) What is the equilibrium concentration of benzene in butter? How much benzene did you buy together with the butter (maximally, $=$ equilibrium)? Log K_{OW} benzene $= 2.1$; $K_{AW} = 0.23$; butter is similar to octanol.

b) How much benzene did you inhale during the 12 minutes you were at the gasoline station (inhalation 12 min = 0.2 m^3)?

c) The average concentration of benzene in urban air is up to 10 μg/m^3. How much benzene do you inhale per day? Compare to the values of 8.4a and 8.4b. Breath volume for one day is 20 m^3.

Chapter 9:
Uptake by Plants

9.1
Significance of the Problem

There are several ways for terrestrial plants to be contaminated with organic xenobiotics. Uptake may be from polluted soils, e.g. former waste dumps or industrial sites. Another risk is uptake from air. Leaves, in particular, are exposed to airborne pollutants.

In some cases, the uptake of toxic substances is desired. Systemic pesticides can only unfold their mode of action when they reach their target within the plant.

The uptake of toxic or potentially harmful substances does not only endanger vegetation. The biomass of terrestrial ecosystems is made up to more than 99.9% by plants. The uptake of chemicals by plants is the first step towards accumulation in the terrestrial food chain and can also be harmful to man.

Nearly all our food – except fish and other aquatic organisms – comes from the terrestrial food web; all grains, fruit, vegetables and meat. In fact, man is an integral part of the terrestrial ecosystem.

It is interesting to compare the legal standards for pollutants in water and food. Table 9.1 shows legal standards for some chemicals in Germany (or the European Union) and the USA. Most values are for pesticides, because legal standards in both water and food could rarely be found for other compounds.

In the European Union, a precautionary standard of 0.0001 mg/L (sum 0.0005 mg/L) for pesticides in drinking water was set. Values in food are toxicologically justified.

This leads to the paradoxical situation in which values in food are up to 1 000 times higher than in drinking water. This also holds for the USA. It is therefore likely that most contaminants make their way to man preferably via food and not via drinking water.

Table 9.1. Legal standards (BBA 1989, BGBl 1986, Rippen 1993; American values: thanks to *J. C. Mc Farlane*)

	water mg/L		meat mg/kg		plants mg/kg	
	EU[a]	US	FRG	US	FRG	US
CCl$_4$	0.003	0.005	nd	nd	0.01–0.1	
Aldicarb	0.0001	0.003	0.01	0.01	0.1–0.5	0.05–1.0
DDT	0.0001	nd	1.0[b]	5.0 (Fish)	0.05–1.0	0.1–3.0
Endosulfane	0.0001	nd	0.1[b]	0.2	0.1–30[c]	0.1–2.0
HCB	0.0001	0.001	0.2[b]	0.3	0.01–0.1	0.05–1.0
Lindane	0.0001	0.0002	1.0–2.0[b]	4.0–7.0	0.1–2.0	0.01–3.0
Mirex	0.0001	nd	0.1[b]	0.1	0.01	nd
Simazine	0.0001	0.004	0.1	0.02	0.1–1.0	0.2–12.0

nd: no data
a) EU: maximal concentration of pesticides in drinking water is 0.0001 mg/L; sum is max. 0.0005 mg/L
b) in fat; c) maize 0.2 mg/kg, fruit and vegetables 1.0 mg/kg, hops 10 mg/kg, tea 30 mg/kg.

9.2
Anatomical and Physiological Principles of Plants

Nearly all terrestrial plants have a similar basic structure. Although the variability is manifold, all constist of roots, stems and leaves. The functions of these plant organs are similar for the species. *Spermatophyta*, with about 250 000 known species the largest and most important class within the plant system, have internal conductive systems for water (the transpir7ation stream within the xylem) and assimilates (within the phloem) (Mauseth 1988). The schematic structure of a plant is shown in Fig. 9.1.

Roots are necessary for anchoring; they take up water and minerals dissolved within. For this purpose, roots have an extremely high surface area. A four-month-old rye plant, for example, has a total root length of 10 000 km with a 1 000 m^2 surface area (Jacob et al. 1987). Since diffusion is proportional to the area, the diffusive exchange between root and soil must be significant.

Water and solutes move freely from soil to the interior of the roots in the capillary spaces between the cortex cells (apparent free space). At the endodermis, this movement is stopped by a barrier of waxy material, the Casparian strip. At the endodermis, water and solutes must pass at least one cell to enter the symplast (Mc Farlane 1995). The membrane is semi-permeable and discriminates some molecules, thus acting like a selective filter controlling the composition of the transpiration stream. Nutrients are taken up actively (by enzymatic action), but for all but a few xenobiotics there is no evidence of active uptake: they pass through the membranes by passive diffusion.

The apoplastic xylem, the conductive system for the transpiration stream, is located in the central cylinder. Water and substances dissolved within are

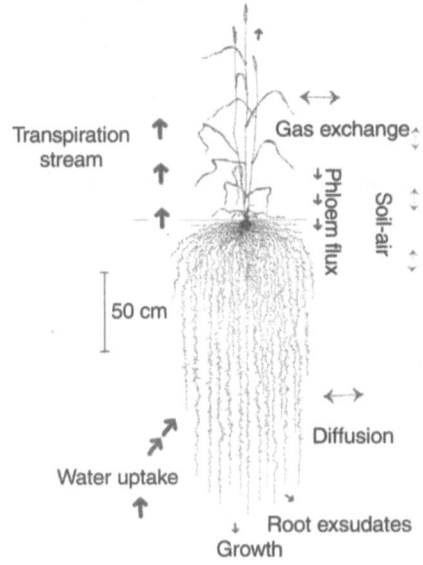

Fig. 9.1. Schematic structure of a plant with conductive systems

transported upwards to the leaves. Depending on the transport conditions and the anatomy of the plant, the flow velocity is up to 150 meters per hour (Huber 1956). For the production of one kilogram of dry biomass, plants in temperate zones transpire between 300 and 650 liters of water. Within one growth period, this is about fifty times the fresh weight of a plant.

Water is transpired from the leaves, whose main function is photosynthesis. Therefore, leaves take up carbon dioxide from air and sunlight. This is why they have a large surface to volume ratio, at least in temperate zones. The epidermis of the leaves on the upper surface differs from the underside. It characteristically has a thicker cuticle and a waxy layer to avoid transpiration and to protect from ultraviolet light. Stomata allowing gas exchange are usually situated in a larger number on the underside. Stomata are closed in the case of water stress and at night.

The assimilates produced in the leaves are transported to the sinks (all growing parts of the plant, storage and fruits) within the phloem (Mauseth 1988). The conductive system of the phloem leads to all parts of the plant. The phloem consists of living cells (symplast) and is usually also situated in the central cylinder. The transport velocity of the phloem is up to 180 cm per hour (Huber 1956). The flux of water in the phloem is approximately ten to hundred times smaller than in the xylem.

The plant lives by interacting with the soil and air environment. Xenobiotics are affected by this. The behavior of xenobiotics within the soil-air-vegetation system may vary considerably. A generic one-compartment box model for the estimation of simultaneous uptake from soil and air is described here (Trapp and Matthies 1995).

9.3
PLANT Model

We assume that anthropogenic organic chemicals are only taken up passively, i.e. by diffusion and advection.

Processes considered are:

- uptake ino roots
- translocation to shoots
- gaseous deposition on leaves
- volatilization from leaves
- metabolism and degradation processes
- dilution by exponential growth.

Partition coefficients for plant tissue

The partition coefficient between plant tissue and water or air is a key property for the fate of compounds in the soil-plant-air-system. For plant tissue, it can be calculated from Eq. 4.12:

$$K_{PW} = (W_P + L_P \, a \, K_{OW}^b)\rho_P/\rho_W \tag{9.1}$$

where K_{PW} is the partition coefficient between plant tissue and water (kg substance/m^3 plant: kg substance/m^3 water), W_P and L_P are the water and lipid content of the plant tissue (g/g), ρ_P and ρ_W are the densities of plant tissue and water (kg/m^3), respectively. 'a' is a correction factor and b is a correction exponent for differences between plant lipids and octanol. For density correction, the value of 'a' is $\rho_{water}/\rho_{octanol} = 1.22$. The exponent b for cut bean roots and stems was found to be 0.75 (Trapp and Pussemier 1991), for macerated barley roots 0.77 (Briggs et al. 1982). For barley shoots, 0.95 was found (Briggs et al. 1983), and 0.97 for isolated citrus cuticles (Kerler and Schönherr 1988).

Uptake into roots

For fine roots diffusive exchange with the soil is high, and near equilibrium conditions are assumed to be achieved. For thicker roots equilibrium is an upper limit, and the kinetics of uptake controls the concentration. The partition coefficient between roots and bulk soil K_{RB} is found by dividing the partition coefficient roots to water (Eq. 9.1 with $b = 0.77$) by the partition coefficient soil to water (eqs. 4.7, 4.10 and 4.11ab, neglecting gas pores):

$$K_{RB} = K_{RW}/(\rho_B K_d + \theta) \tag{9.2}$$

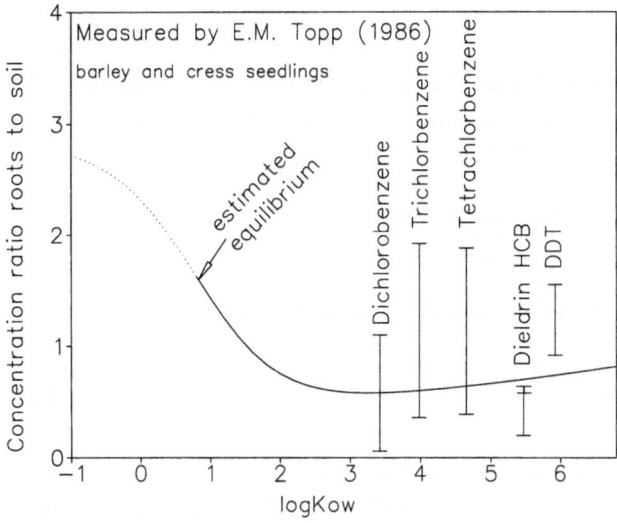

Fig. 9.2. Calculated and measured equilibrium between roots and soil

where K_d is the distribution coefficient between soil matrix and water (L/kg), ρ_B is the bulk soil density (here: kg/L) and θ is the volumetric water fraction of the soil. K_{RW} is the partition coefficient between roots and water (mass/vol. to mass/vol.), calculated analogously to Eq. 9.1. Values of Eq. 9.2 are close to one and only slightly depend on lipophilicity when K_d is calculated from K_{OC} using Eq. 4.11b, the method of Schwarzenbach and Westall (1981). A plausible reason for the similar sorption properties of roots and soil is that the humic substances in soil originate from plant material.

Measurements by E. M. Topp for barley and cress roots indeed support this finding (Fig. 9.2). Schroll and Scheunert (1992) found K_{RW}-values of hexachlorobenzene for maize, oats, rape and barley between 0.8 and 3.2, but higher values for lettuce (13.7) and carrots (31.6).

It could be concluded that it is sufficient to assume that the concentration in fine roots is usually in or below the order of magnitude of the concentration in soil, with some exceptions.

When acid-base reactions occur, an additional process, the 'ion trap', may also occur (pH in phloem is approximately 8, 5.5 in xylem; relatively independent from soil pH; 'ion trap' see Briggs et al. 1987 and Rigitano et al. 1987).

Translocation with the transpiration stream

Roots take up a lot of water and by this substances dissolved within the water. The 'Transpiration Stream Concentration Factor' *TSCF* is defined as the concentration ratio between xylem sap and external solution (soil water). The mass transport to stem and leaves within the xylem N_{Xy} (kg/s) is then

$$N_{Xy} = Q\, C_W\, \mathit{TSCF} \tag{9.3}$$

where Q is the transpiration stream (m^3/s), C_W is the concentration in soil water, C_W is C_B/K_d and C_B is the concentration in (dry) bulk soil. The TSCF is related to the K_{OW} (Briggs et al. 1982):

$$\mathit{TSCF} = 0.784 \exp \frac{-(\log K_{OW} - 1.78)^2}{2.44} \tag{9.4a}$$

Hsu et al. (1990) found an equation of a similar form:

$$\mathit{TSCF} = 0.7 \exp \frac{-(\log K_{OW} - 3.07)^2}{2.78} \tag{9.4a}$$

Hsu et al. used a pressure chamber technique. This gives faster xylem fluxes and less time for equilibration. Maybe, this explains the higher $\log K_{OW}$ optimum of their TSCF-equation.

From the comparison of both empirical equations it can be seen that the TSCF is an uncertain parameter, in particular, for very lipophilic substances. Measured values have a large variance (Fig. 9.3). We propose the use of the higher result from Eqs. 4a and 4b. From theoretical considerations it follows that TSCF-values of non-dissociating compounds should maximally be 1.

An accumulation of chemical in the shoot and, in particular, in foliage may then occur when the substance only volatilizes and metabolizes slowly.

Note: The calculation of the mass balance in aerial plant parts is independent of the roots, because the TSCF is directly related to the soil solution.

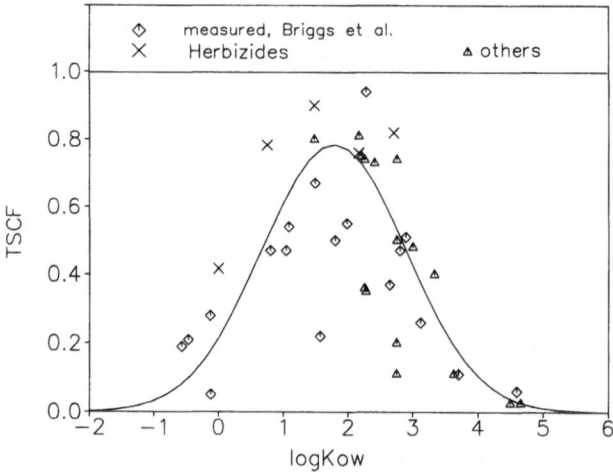

Fig. 9.3. TSCF versus $\log K_{OW}$; adapted from Briggs et al (1982)

Exchange with air

The partition coefficient between leaves and air K_{LA} is

$$K_{LA} = K_{LW}/K_{AW} \tag{9.5}$$

where K_{AW} is the partition coefficient between air and water (dimensionless Henry's law constant) and K_{LW} is the partition coefficient between leaves and water (dimensionless), calculated from Eq. 9.1. The coefficient b is likely to be plant specific. We use experimental data from Briggs et al. (1983) for barley shoots ($b = 0.95$). Values of K_{LW} are usually $\gg 1$, of $K_{AW} \ll 1$; therefore, K_{LA} can have very high values (see Fig. 9.4). This implies that an uptake and accumulation of organic vapors from air into leaves can occur, even if aerial concentrations are very small.

Note: K_{LW} is similar to SXCF ρ_L/ρ_W, where SXCF is the stem xylem concentration factor of Briggs et al. (1983).

The diffusive net flux between leaves and atmosphere ('gaseous dry deposition') N_A (kg/s) is:

$$N_A = A\,g\,(C_A - C_L/K_{LA}) \tag{9.6}$$

where A is the leaf surface area (m^2), g is the conductance (m/s), C_A is the gas phase concentration in air (kg/m^3) and C_L is the concentration in leaves (kg/m^3). Estimates of average values for conductance g (m/s) are:

lower boundary: cuticle is comparatively impermeable; uptake is mainly via stomata (vapors, approx. when log K_{OW} − log K_{AW} < 5); conductance g is approx. 0.001 to 0.0001 m/s, depending on plant species and environmental conditions.

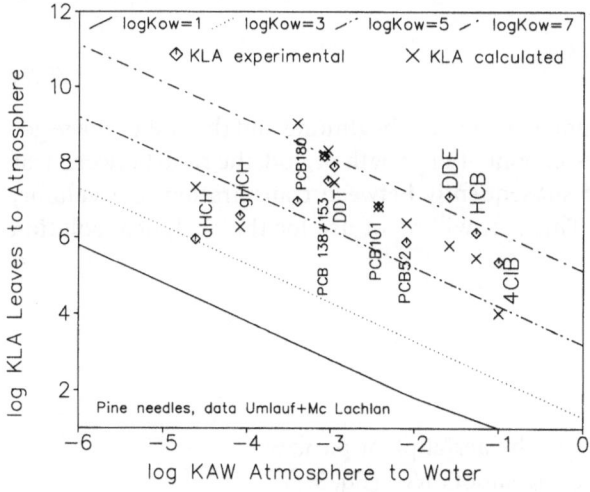

Fig. 9.4. Partition coefficient between leaves and air

upper boundary: cuticle is relatively permeable (very lipophilic compounds, approx. when $\log K_{OW} - \log K_{AW} > 10$); the main resistance is from the atmospheric boundary layer, g is approx. 0.005 m/s, Thompson (1983).

A default value of $g = 0.001$ m/s is proposed. For details see Riederer (1995).

The PLANT model in CemoS enables the user to calculate stomatal and cuticular resistances for specific chemical, plant and environmental conditions. For further details refer to the manual or Trapp (1995).

The model also optionally enables the user to calculate dry and wet deposition of particles to leaf surfaces. The methodology is similar to the deposition processes in AIR, please refer to Chapter Eight of the manual.

Metabolism and photodegradation

Plant metabolism differs from that of animals, since plants have no excretion organ and bound residues are often formed (Komoßa et al. 1995). Photodegradation is likely to occur, since leaves are exposed to sunlight. Pseudo first-order kinetics are assumed for the calculations. Rate constants for degradation in plants are not calculated, but must be supplied from experiments, literature or external estimation routines. $\lambda_E(\mathrm{d}^{-1})$ is the elimination rate constant, including metabolism and photodegradation.

Growth

The growth of plants depends on the stage of development. Shortly after germination, growth is small. Then the vegetative phase follows during which the main growth occurs. Finally, during the maturation stage, the growth comes to an end. During the vegetative phase, growth can be approximated by an exponential function, and the dilution by growth can be calculated by a growth rate $\lambda_G(\mathrm{d}^{-1})$:

$$\lambda_G = \ln(V_{\mathrm{end}}/V_0)/t \tag{9.7}$$

where V_0 and V_{end} are the volumes (m^3) at the beginning and the end of the vegetation period $t(d)$. During the exponential growth period, the ratio between leaf area and volume, A/V_L, and subsequently between transpiration and volume, Q/V_L, is relatively constant. This fact will be of use for the analytical solution (see below).

Mass balance

The mass balance is:

Change of chemicals mass in the aerial plant parts =
+ flux from soil via xylem to the shoots N_{Xy} (Eq. 9.3)

± gaseous flux from/to air N_A (Eq. 9.6)
− photodegradation − metabolism

Expressed in mathematical terms:

$$dm_L/dt = d(C_L V_L)/dt = Q \text{ TSCF } C_W + Ag(C_A - C_L/K_{LA}) - \lambda_E m_L \tag{9.8}$$

where m_L and V_L are mass and volume of the leaves (m³) (and functions of time t). When growth is exponential and ratios A/V_L and Q/V_L are assumed to be constant, it then follows for the change of the concentration with time dC_L/dt

$$dC_L/dt = -[Ag/(K_{LA}V_L) + \lambda_E + \lambda_G]C_L + C_W \text{ TSCF } Q/V_L + C_A gA/V_L \tag{9.9}$$

Analytical solution for constant conditions

Taking the parameters on the right side of Eq. (9.9) as constants yields a linear differential equation of first order:

$$dC_L/dt = -aC_L + b \tag{9.10}$$

where $a = Ag/(K_{LA}V_L) + \lambda_E + \lambda_G$ (sink terms)
and $b = C_W \text{ TSCF } Q/V_L + C_A gA/V_L$ (source terms).

With a given $C_L(0)$ the analytical solution of the equation is:

$$C_L(t) = C_L(0) \exp(-at) + b/a[1 - \exp(-at)] \tag{9.11}$$

The steady-state concentration ($t \rightarrow \infty$, $dC_L/dt \rightarrow 0$) is

$$C_L(\infty) = b/a \tag{9.12}$$

The time to reach steady-state (95%) is:

$$t(95\%) = -\ln 0.05/a \tag{9.13}$$

9.4
Limitations

The complex dynamic behavior of a chemical in the soil-plant-air system is reduced to one equation. This implies limitations.

The approach is developed for non-ionic organic substances. For the extension to dissociating chemicals, see Briggs et al. (1987). Fertilizers and inorganic, dissociating compounds are taken up by plant transport systems; their behavior is different.

There is no spatial differentiation of the plant. Calculated concentrations correspond to the aerial plant compartment, mainly foliage. Concentrations in fruit can deviate largely. Exponential growth is assumed. This is only valid for plants that are harvested before maturation, e.g. green fodder, green vegetables and lettuce.

Transport of chemicals within the phloem is usually much smaller than within the xylem, but for chemicals with specific properties (e.g. some dissociating chemicals), it becomes very important (Bromilow and Chamberlain 1995).

It is not clear whether deposition by aerosols contributes significantly to accumulation from air. At present, in the PLANT model in CemoS the calculation of deposition of particles is optional.

The soil-air-plant path which is of significance for volatile and semivolatile lipophilic chemical needs a different solution (Trapp and Matthies 1997).

The empiric parameters used in this equation (K_{LA}, TSCF) were derived from a small number of experiments. The TSCF of nitrobenzene for seven plant species differed less than 10% (Mc Farlane et al. 1990) and was close to the results of eqs. 4a and 4b. But no experimental data are available for chemicals with high log K_{OW}.

The given parametrization is not for a specific plant, but represents average values. Moreover, properties are taken as constant. This is convenient for the mathematical solution of the equation, but is not very realistic and can lead to error.

It becomes clear from these limitations (the reader may find more) that the equation is of a generic type and less applicable for real situations, where complex numerical models might be advantageous.

Example 9.1: Application for the uptake of 2,3,7,8-TCDD

Dioxins and related compounds are a highly toxic class of compounds. A major source for 2,3,7,8-TCDD is the uptake by food, e.g. milk and meat. Grazing animals take up PCDD/F from the grass they feed on. But how does 2,3,7,8-TCDD enter the plants? The following calculation addresses this question (Trapp and Matthies 1995). Input data for the calculation are given in Table 9.2. Plant and soil values are typical for a pasture. They are normalized to one square meter. The concentrations of 2,3,7,8-TCDD in soil and air are from measurements at a rural site (McLachlan 1992). A laboratory experiment was carried out for the determination of the photodegradation of 2,3,7,8-TCDD (McCrady and Maggerd 1993). The resulting rate λ_E (Table 9.2) is multiplied by 0.3 (30% of the time full sunlight). It is not clear whether this rate constant is valid for environmental conditions. In an additional sensitivity analysis, λ_E is set to zero.

Example of calculation

1. Convert all units to SI (m, kg, s), except chemicals mass; $1000\,\text{fg} = 1\,\text{pg} = 10^{-15}\,\text{kg}$
2. Concentration in soil solution C_W

$$
\begin{aligned}
C_W \approx C_B/K_d &= C_B/(OC\,K_{OC}) \\
&= 70\,\text{pg/kg}/(0.01 \cdot 2\,365\,058\,\text{L/kg}) \\
&= 0.003\,\text{pg/L} = 3\,\text{pg/m}^3
\end{aligned}
$$

3. Partition coefficient leaves to atmosphere K_{LA}

$$
\begin{aligned}
K_{LW} &= (W_P + L_P\,a\,K_{OW}^b)\rho_P/\rho_W \\
&= \{0.8 + 0.02 \cdot 1.22 \cdot (10^{6.76})^{0.95}\} \cdot 1/2 \\
&= 32\,237
\end{aligned}
$$

$$
K_{LA} = K_{LW}/K_{AW} = 2.1 \cdot 10^7
$$

4. Calculation of the $TSCF$

(9.4a) $TSCF = 0.784 \exp[-(6.76 - 1.78)^2/2.44] = 3.0 \cdot 10^{-5}$
(9.4b) $TSCF = 0.7 \exp[-(6.76 - 3.07)^2/2.78] = 5.2 \cdot 10^{-3}$

higher $TSCF$ is $5.2 \cdot 10^{-3}$

Table 9.2. Input data for the example calculation

Environment (normalized to 1 m^2)		
Shoot mass	1 kg	
Shoot volume V	0.002 m^3	
Leaf area A	5 m^2	
Lipid content L_P	0.02 g/g	
Water content W_P	0.8 g/g	
Transpiration Q	1 L/d = $1.15 \cdot 10^{-8}$ m^3/s	
time to harvest	60 days	
Growth rate constant λ_G	0.035 d^{-1}	
Soil organic carbon content OC	1%	
Bulk soil density ρ_B	1.3 kg/L	
Data 2,3,7,8-TCDD		
log K_{OW}	6.76	(a)
log K_{OC}	6.37	calculated, (c)
K_{AW}	0.0015	(a)
rate constant λ_P	0.3744 d^{-1}	
C_B (dry soil matrix)	70 pg/kg	(b)
C_A (air, gaseous)	2.7 fg/m^3	(b)

a) Rippen (1996)
b) McLachlan (1992)
c) using the equation of Karickhoff, Eq. 4.11a

5. Uptake term $b = Q\ TSCF\ C_W/V_L + g\ A\ C_A/V_L$

    uptake from soil b_1

    $$
    \begin{aligned}
    b_1 &= 1.15 \cdot 10^{-8}\,\mathrm{m^3/s} \cdot 5.2 \cdot 10^{-3} \cdot 3\,\mathrm{pg/m^3}/0.002\,\mathrm{m^3} \\
        &= 9 \cdot 10^{-8}\,\mathrm{pg\,s^{-1}m^{-3}} = 7.7 \cdot 10^{-3}\,\mathrm{pg\,d^{-1}m^{-3}}
    \end{aligned}
    $$

    uptake from air, gaseous b_2

    $$
    \begin{aligned}
    b_2 &= A\,g\,C_A/V_L = 5\,\mathrm{m^2} \cdot 10^{-3}\,\mathrm{m/s} \cdot 2.7\,\mathrm{fg/m^3}/0.002\,\mathrm{m^3} \\
        &= 6.75\,\mathrm{fg\,s^{-1}m^{-3}} = 583\,\mathrm{pg\,d^{-1}m^{-3}}
    \end{aligned}
    $$

    total uptake term b is

    $$
    b = b_1 + b_2 = 6.75\,\mathrm{fg\,s^{-1}m^{-3}} = 583\,\mathrm{pg\,d^{-1}m^{-3}}
    $$

6. Sink term $a = A\,g/(K_{LA}V_L) + \lambda_E + \lambda_G = a_1 + a_2 + a_3$
    a_1 is loss by volatilization from leaves to air

    $$
    \begin{aligned}
    a_1 &= 5\,\mathrm{m^2} \cdot 10^{-3}\,\mathrm{m/s}/(2.1 \cdot 10^7 \cdot 0.002\mathrm{m^3}) \\
        &= 1.2 \cdot 10^{-7}\,\mathrm{s^{-1}}
    \end{aligned}
    $$

a_2 is loss by photodegradation

    $$
    a_2 = 0.3744\,\mathrm{d^{-1}} \cdot 0.30 = 1.11 \cdot 10^{-6}\,\mathrm{s^{-1}}
    $$

a_3 is dilution

    $$
    \begin{aligned}
    a_3 &= 0.035\,\mathrm{d^{-1}} = 4.0 \cdot 10^{-7}\,\mathrm{s^{-1}} \\
    a   &= a_1 + a_2 + a_3 = 1.63 \cdot 10^{-6}\,\mathrm{s^{-1}}
    \end{aligned}
    $$

7. Time to reach steady-state (95%):

    $$
    \begin{aligned}
    t(95\%) &= -\ln 0.05/a = 3/(1.63 \cdot 10^{-6}\,\mathrm{s^{-1}}) \\
            &= 1.84 \cdot 10^6\,\mathrm{s} = 21\,\mathrm{d}
    \end{aligned}
    $$

8. Steady-state (95%) is reached within 60 days

    $$
    \begin{aligned}
    C_L(\infty) &= b/a = 6.75 \cdot 10^{-3}\,\mathrm{pg\,s^{-1}m^{-3}}/1.63 \cdot 10^{-6}\,\mathrm{s^{-1}} \\
                &= 4.14\,\mathrm{ng\,m^{-3}} = 8.3\,\mathrm{pg/kg}\ (\text{wet wt.})
    \end{aligned}
    $$

9. Sensitivity analysis: calculation without photodegradation
    loss term $a = 5.2 \cdot 10^{-7}\,\mathrm{s^{-1}}$

    $$
    \begin{aligned}
    t(95\%) &= -\ln 0.05/a = 3/5.2 \cdot 10^{-7}\,\mathrm{s^{-1}} = 66.7\,\mathrm{d} \\
    C_L(\infty) &= b/a = 6.75\,\mathrm{fg\,s^{-1}m^{-3}}/(5.2 \cdot 10^{-7}\,\mathrm{s^{-1}}) \\
                &= 13\,\mathrm{ng\,m^{-3}} = 26\,\mathrm{pg/kg}(\text{wet wt.}) \\
    C_L(60d) &= b/a[1 - exp(-a \cdot 60 \cdot 86400] \\
             &= 24.2\,\mathrm{pg/kg}\ (\text{wet wt.})
    \end{aligned}
    $$

From the calculation it can be concluded that for background concentrations, uptake of 2,3,7,8-TCDD into foliage is from air. Calculated 2,3,7,8-TCDD concentrations in plants are close to those measured in grass at Bayreuth (McLachlan 1992), 30 to 50 pg/kg dry weight, which is 6 to 10 pg/kg wet wt., and those from a rural site at Rothamstad, UK, (32 pg/kg dry wt.)(Kjeller et al. 1991).

Exercises on chapter nine

9.1 Uptake from soil

Nitrobenzene is a product resulting from the use of TNT. It is often found on former military training areas. Let the concentration of nitrobenzene ($\log K_{OW} = 1.85$) in the dry soil matrix be 5 ppm (5 mg/kg dry soil).

Data: $K_d = 0.58\,cm^3$ (soil solution)/g (dry soil);
lettuce: transpiration 50 liters in 50 days; harvest after 50 days; harvest weight 1 kg.

a) What is the concentration in the soil solution?
b) What amount of nitrobenzene is translocated into the (leaves of) the lettuce?

9.2 Uptake from air
Assume an average concentration of nitrobenzene in the air of $3\,\mu g/m^3$.

a) What would be the equilibrium concentration to air of nitrobenzene in lettuce?
   additional data: lipid content 2%, water content 80%, K_{AW} 0.00061, density 1 kg/L.
b) What is the concentration in initially clean lettuce following uptake from air after 1 hour, 50 days (harvest)?

additional data: conductance $g = 3 \cdot 10^{-4}\,m/s$, leaf area $A = 2\,m^2$, growth rate $\lambda_w = 0.23\,d^{-1} = 2.66 \cdot 10^{-6}\,s^{-1}$, metabolism half-life $= 1.44$ days, $\lambda_E = 0.48\,d^{-1} = 5.6 \cdot 10^{-6}\,s^{-1}$.

9.3 Competitive uptake from soil and air
With the given data calculate the simultaneous uptake from soil and air for the steady-state. Which way does nitrobenzene take, and why? How long does it take until steady-state is reached (95%)?

Chapter 10:
Model for Food Chains

10.1
Problem Definition

Persistent, lipophilic chemicals accumulate in the food chain. Typical food chains are:

algae → daphnia → small fish → predatory fish → predatory bird
vegetation → rabbit → hawk
vegetation → cow → milk → (mother → milk) → baby

Predators, humans, and in particular babies, are at the top of the food chain. Indeed, mother's milk is in many cases more contaminated than food (e.g. with chloroorganics).

10.2
Mathematical Formulation of the CHAIN Model

The food chain can be considered analogously to the cascade in Chapter Two. In the first trophic level of the food chain, a part of the pollutant's mass is metabolized and eliminated, and a part is eaten by the next trophic level. The metabolism plus elimination rate is λ_1, the rate of transfer from trophic level one to level two is k_{12}. The mass balance of the chemical in the first trophic level is

$$dm_1(t)/dt = -\lambda_1 m_1(t) - k_{12}m_1(t) \tag{10.1}$$

For trophic level two, there is uptake by transfer from the lower level and loss by transfer to the higher level three plus loss by metabolism and elimination in trophic level two:

$$dm_2(t)/dt = k_{12}m_1(t) - \lambda_2 m_2(t) - k_{23}m_2(t) \tag{10.2}$$

We assume that the third is the highest trophic level and no transfer to higher levels occurs. Remaining processes are uptake and metabolism plus elimination:

$$dm_3(t)/dt = k_{23}m_2(t) - \lambda_3 m_3(t) \tag{10.3}$$

The matrix is

$$dm(t)/dt = \left\{ \begin{array}{ccc} -k_{12} - \lambda_1 & 0 & 0 \\ k_{12} & -k_{23} - \lambda_2 & 0 \\ 0 & k_{23} & -\lambda_3 \end{array} \right\} m(t)$$

The solution is analogous to example 2.3 (see also the exercises on this Chapter). With the initial conditions $m_1(0) = m_0$, $m_2(0) = m_3(0) = 0$ it follows

$$m_1(t) = m_0 \exp(-k_{12}t - \lambda_1 t) \tag{10.4}$$

$$m_2(t) = \frac{m_0 k_{12}}{\lambda_2 + k_{23} - \lambda_1 - k_{12}}$$
$$\cdot \left\{ \exp(-k_{12}t - \lambda_1 t) - \exp(-k_{23}t - \lambda_2 t) \right\} \tag{10.5}$$

$$m_3(t) = k_{12}k_{23}m_0 \left\{ \frac{\exp(-k_{12}t - \lambda_1 t)}{(\lambda_2 + k_{23} - \lambda_1 - k_{12})(\lambda_3 - \lambda_1 - k_{12})} \right.$$
$$+ \frac{\exp(-k_{23}t - \lambda_2 t)}{(\lambda_2 + k_{23} - \lambda_1 - k_{12})(\lambda_2 + k_{23} - \lambda_3)}$$
$$\left. + \frac{\exp(-\lambda_3 t)}{(\lambda_3 - \lambda_2 - k_{23})(\lambda_3 - \lambda_1 - k_{12})} \right\} \tag{10.6}$$

Example 10.1: 2,3,7,8-TCDD in the food chain

The 'Seveso-dioxin' 2,3,7,8-TCDD is an example of a persistent substance that is transferred in the food chain. We consider the pasture-cow-milk path.

a) *The food chain*:
The pasture has an area of one hectare and has a fresh weight of 1 kg per m². The growth (in summer) is 0.3 kg/m² in 60 days or 50 kg/(ha d).

Xarne, the cow, is grazing on it. It feeds 50 kg/d, the total growth. Xarne provides 25 liters of milk per day (25 kg/d).

b) *The input*:
The pasture is close to a chemical plant. 2,3,7,8-TCDD is released due to an accident. 100 µg are blown on the pasture, which gives 10 ng/kg grass.

c) *The substance behavior* (assumptions):
2,3,7,8-TCDD is 'eliminated' (volatilizing, photodegraded, washed off) with a half-life of two weeks or at a rate of 0.05 d⁻¹. The cow resorbs 70% (McLachlan 1992) and the rest is redeposited on the pasture. 2,3,7,8-TCDD is persistent in the cow. The clearance half-life via milk is 40 days (McLachlan 1992). Thus, the transfer rate from the cow to the milk k_{23} is 0.017 d⁻¹. There is no degradation

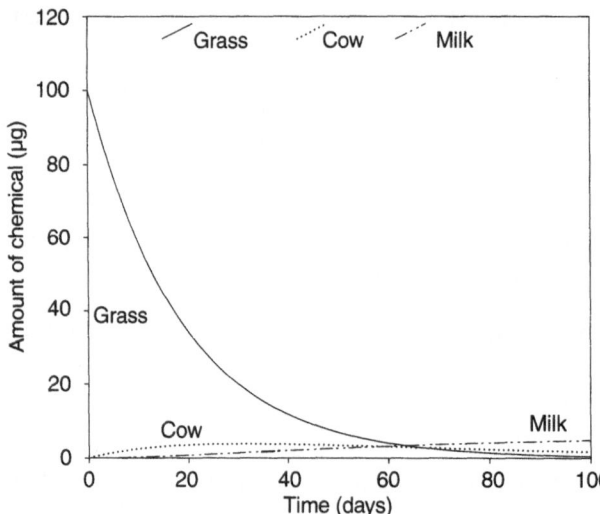

Fig. 10.1. Mass of 2,3,7,8-TCDD in the food chain

in the milk. For the calculation we assume constant transfer and degradation rates.

d) *The balance equation*: Input data for Eqs. 10.4 to 10.6 are:

$m_0 = 100\,\mu g$; $\lambda_1 = 0.05\,d^{-1}$;
$k_{12} = 0.7 \cdot 50\,kg/10000\,kg = 0.0035\,d^{-1}$;
$\lambda_2 = 0$; $k_{23} = \ln 2/40\,d = 0.017\,d^{-1}$; $\lambda_3 = 0$.

d) From mass to concentration
The concentration results from the division of mass by volume. The volume of the pasture is constant, we assume that the cow feeds as fast as the grass grows. The volume of the adult cow is constant, too.

Concentration in the grass: $m_1/10\,000\,kg$
Concentration in the cow: $m_2/500\,kg$

The milk volume increases linearily every day. The concentration in milk for a given day is

$$C_3(t) = \{m_3(t) - m_3(t - \Delta t)\} / \{V_3(t) - V_3(t - \Delta t)\}$$

Δt : usually one day, corresponding volume is 25 liters.

The results of the calculations can be seen in Figs. 10.1 and 10.2.

Most of the chemical's mass is found in the grass during the first month. Later, the mass is highest in the milk, because the absolute milk volume increases steadily. The concentration of 2,3,7,8-TCDD is at first highest in the grass, but after one month far higher in the cow and the milk. While the concentration in the grass decreases quickly, the persistent dioxin is still found

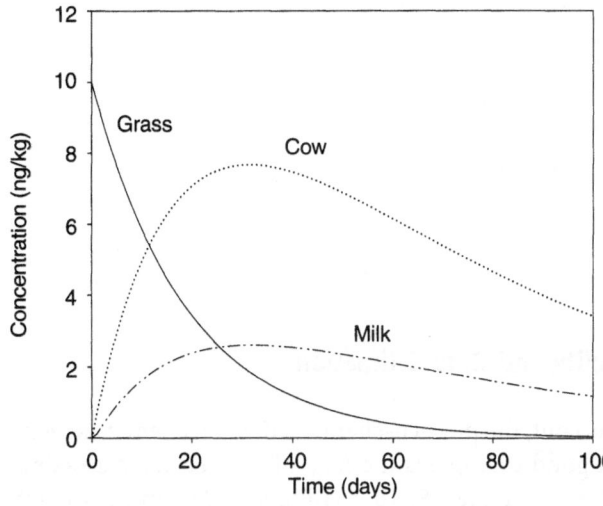

Fig. 10.2. Concentration of 2,3,7,8-TCDD in the food chain

in the food chain for a long time. This is why investigations of milk samples directly after an accident are not very useful and have to be repeated after several days and weeks.

Exercises on chapter ten

1) Calculate masses and concentrations of 2,3,7,8-TCDD in Xarne's meat and milk 60 days after the accident, if there is no elimination in grass ($\lambda_1 = 0$).
2) Apply the general solution scheme for linear differential equation systems (Eq. 2.27) to the cascade example in Chapter Two (example 2.3).

The system for $n = 3$ is:

$$
\begin{aligned}
dm_1(t)/dt &= -k_1 m_1(t) \\
dm_2(t)/dt &= +k_1 m_1(t) - k_2 m_2(t) \\
dm_3(t)/dt &= +k_2 m_2(t) - k_3 m_3(t)
\end{aligned}
$$

Chapter 11:
Data Estimation

11.1
Some Words on Data Quality and Data Estimation

As the careful reader will find out, the physico-chemical data on chemicals mentioned in this book vary. A good example is the K_{AW} of the herbizide atrazine. In the database (Section 11.8) we wrote $7.9 \cdot 10^{-8}$ (Rippen 1996). In the CemoS data base, we have the similar value $7.86 \cdot 10^{-8}$. The latter one is the estimated value that follows from water solubility and vapor pressure. In Table 4.1 we give the value $8.05 \cdot 10^{-9}$ (Nirmalakhandan and Speece 1988). This is not to confuse our readers. We as modellers depend on those data, and we cannot decide ourselves which ones are 'good' or 'bad'. So we have a problem. Variations of factor 10 or more can significantly change the results of some simulations. This does not mean that the simulation is wrong, but it is uncertain.

One way to solve this problem is to use estimation routines and compare the values estimated with those ones experimentally determined. Some obviously wrong data may be eliminated by this way. Often it is difficult to decide which data is "better". This uncertainty can be handled by an uncertainty analysis, this is to run the model with different combinations of data sets and see how the results vary. Often, this is done by a so-called 'Monte Carlo' analysis: the input data are stochastically varied and the output of the model is a distribution of possible results. As long as data are uncertain (they always are), the outcome of simulations is a probability distribution rather than one distinct value. We will not discuss this here in detail, because it is a more general subject and not specific for environmental modeling. However, stochastic variations of environmental data are high (e.g. in space and time). Physico-chemical data are more uncertain for extreme ranges (e.g. very low vapor pressure $\rightarrow$ very low K_{AW}; very small water solubility $\rightarrow$ very high K_{OW}), because the measurement is far more difficult.

We have included the following features into CemoS to avoid some errors:
– All data can/must be commented with an own comment field.
– Via the MIF (Model Interchange Format), all data plus all comments are exported.

- Ranges of plausibility and ranges of possibility are given and checked. Physically impossible values are not accepted.
- 'Closed Loop estimation' is not possible (e.g. you cannot estimate K_{OW} from K_{OC} and then estimate K_{OC} from K_{OW} and so on).
- Estimation functions may be used to check whether a value is likely or not.

Although this helps to avoid some errors, many others may still occur. So, check data and results carefully and do not assume the outcome of a model to be an absolute value: Compare to experimental data, perform uncertainty studies, and accept the model results as a help in gaining new knowledge and interpretation of measured concentrations.

Data estimation

The models shown above have similar data requirements: molar mass, the partition coefficients K_{AW} (air to water), K_{OC} (organic carbon to water) and K_{OW} (octanol to water), plus the reaction or metabolism rate constants.

Often the problem arises that no adequate data can be found. Then the estimation of missing substance properties from known ones may help (property-property estimation). Errors may, however, always be introduced by estimation. No estimation functions are given for reaction and transformation rates, because the correlation to physico-chemical properties is often weak. More information can be found within the 'Handbook of Chemical Property Estimation Methods' by Lyman et al. (1990). The following equations 11.1 to 11.14 are offered within the CemoS program.

11.2
Calculation of Molar Mass

The molar mass M (g/mol) can easily be estimated from the chemical formula by adding the masses of the elements (given in the periodic system of elements):

molar mass of chloroform $CHCl_3 = 1 \, C + 1 \, H + 3 \, Cl$
$M = 12.01 \, \text{g/mol} + 1.0 \, \text{g/mol} + 3 \cdot 35.45 \, \text{g/mol} = 119.36 \, \text{g/mol}$

11.3
Partition Coefficient n-Octanol to Water K_{OW}

The partition coefficient octanol to water K_{OW} describes the lipophilicity of a chemical and is one of the key parameters in exposure models. It is negatively correlated to water solubility. Yalkowski and Valvani (1980) found the following regression which also considers the aggregate state at 25°C:

$$\log S_m = -1.05 \cdot \log K_{OW} + 0.87 - 0.012 \cdot f(T_m) \tag{11.1}$$

or:

$$S = M \cdot 10^{[-1.05 \log K_{OW} + 0.87 - 0.012 f(T_m)]}$$

S_m :   water solubility (here: mol/L)
S :   water solubility in g/L or kg/m^3 (SI unit)
T_m :   melting point °C
$f(T_m)$:   if $T_m > 25°C = T_m$; else $= 25°C$;

A comprehensive study showed that this regression fails for some substance classes with OH group (in particular phenols). For the other cases, this regression was superior to 25 others (Brüggemann and Altschuh 1991).

The K_{OW} inversely follows from the water solubility S_m:

$$\log K_{OW} = -0.95 \cdot \log S_m + 0.83 - 0.0114 \cdot f(T_m) \qquad (11.2)$$

or

$$\log K_{OW} = -0.95 \cdot \log(S/M) + 0.83 - 0.0114 \cdot f(T_m)$$

Note: Solubility S_m [mol/L] $\cdot M$ [g/mol] $= S$ [g/L = kg/m^3]

11.4
Partition Coefficient Organic Carbon to Water K_{OC}

The sorption of lipophilic organic compounds to humic substances describes the K_{OC}. It can be estimated from the K_{OW} by the equations of Karickhoff (1981) and Schwarzenbach and Westall (1981) (Eqs. 4.11a and b):

Karickhoff:

$$K_{OC} = 0.411 \, K_{OW} \qquad (11.3)$$

Schwarzenbach and Westall:

$$\log K_{OC} = 0.72 \log K_{OW} + 0.49 \qquad (11.4)$$

K_{OC} is the partition coefficient between organic carbon and water (cm^3/g).
Limitations:
These equations are only valid for non-dissociating organic chemicals, not for ions, acids or bases (see Sect. 4.3). The maximum concentration in the aqueous phase must be below half of the solubility. For both low or high extreme values of K_{OW}, considerable deviations between both equations occur, indicating uncertainty of the estimation.

Karickhoff's equation (11.3 or 4.11a) was found for five polycyclic aromatic hydrocarbons and a number of sediments ($r^2 = 0.994$), the log K_{OW} of the chemicals was between 1.0 and 6.72. The equation of Schwarzenbach and Westall (Eq. 11.4 or 4.11b) was found for a series of alkylated and halogenated benzenes, log K_{OW} ranging from 2.6 to 4.72 and several soils with an organic carbon content OC of 0.04% to 33%. According to the authors, Eq. 11.4 gives results within a factor of two deviating from experimental values, when $OC > 0.1\%$.

If only water solubility is given, the K_{OW} and then the K_{OC} can be estimated (estimation errors increase!). There are also direct relations, e.g. from Kenaga and Goring (1980):

$$\log K_{OC} = 3.64 - 0.55 \log(S \cdot 1000) \qquad N = 106; \; r^2 = 0.71 \qquad (11.5)$$

S: water solubility in kg/m^3 (mg/L in the original work); regression range of S [kg/m^3] is $0.5 \cdot 10^{-6}$ to $1\,000$ (Lyman et al. 1990); a variety of substances, mainly pesticides.

From Chiou et al. (1979) ($N = 22$, $r^2 = 0.933$):

$$\log K_{OC} = -0.686 \log(S \cdot 1000) + 4.273 \qquad (11.6)$$

Water solubility S is in kg/m^3.

11.5
Partition Coefficient Between Air and Water K_{AW}

The partition coefficient between air and water K_{AW}, the so-called dimensionless Henry's law constant, is estimated from the saturation vapor pressure p_s and the water solubility S. The vapor pressure sometimes has obscure units. For conversion to the SI-unit Pascal, use Table 11.1. The partition coefficient is

$$K_{AW} = \frac{H}{RT} = \frac{p_s}{S_m R T} = \frac{M\, p_s}{1000\, S\, R\, T} \qquad (11.7)$$

p_s is the saturation vapor pressure (Pa) (with solids: vapor pressure of the subcooled liquid), S_m is the water solubility of the substance in mol/m^3, and S is measured in kg/m^3. R is the universal gas constant (8.314 J mol^{-1}K^{-1}), and T is temperature (K). M is the molar mass in g/mol (the factor 1/1000 results from converting g to kg).

The equation is valid within the range $S < 1000$ mol/m^3. For substances that are gases at given temperature (boiling point < environmental temperature),

Table 11.1. Units of pressure p

Pascal, Pa	$= $ N/m^2	$=$ kg m s^{-2}	(SI)
Bar, bar	$= 0.1$ Mpa	$= 10^5$ Pa	
techn. atmosphere, at	$= 1$ kp/cm^2	$= 98\,066.5$ Pa	
physical atmosphere, atm	$= 760$ Torr	$= 101\,325$ Pa	
mm mercury column, Torr		$= 133.3224$ Pa	
meter water column, m WS	$= 0.1$ at	$= 9\,806.65$ Pa	
pound-force per square inch, lbf/in^2	$=$ psi	$= 6\,894.76$ Pa	

the atmospheric pressure is used (101 325 Pa) instead of p_s. For substances that are completely mixable with water (e.g. ethyl alcohol), Raoult's law is used instead of Henry's law (see physical chemistry text books, e.g. Atkins 1986). For ions, salts and completely dissociated compounds, $K_{AW} = 0$; for incompletely dissociating compounds see Sect. 4.3.

11.6
Estimation of Bioconcentration Factors for Biota

The bioconcentration factor is the concentration ratio between biomass and its surrounding medium, usually water. Unlike partition coefficients, it is usually given in the unit 'mass of chemical per mass biota to mass of chemical per mass of water'.

Estimation in the PLANT model for leaves, roots and shoots:

The correlation between n-octanol and plant lipids is a solvent-solvent regression equation, the principal equation is

$$K_{SW} = a\, K_{OW}^b \tag{11.8}$$

where K_{SW} is the partition coefficient between solvent and water, K_{OW} that between octanol and water, and a and b are empirical coefficients. In case of plant tissue, the 'solvent' are the plant lipids. This leads to the equation

$$BCF = W + L_P a\, K_{OW}^b \tag{11.8a}$$

W_P : water content (mass per mass fresh weight)
L_P : lipid content (mass per mass fresh weight)
b : exponent to correct differences between n-octanol and plant lipids. In PLANT, b is 0.77 for roots and 0.95 for leaves. The exponents were empirically determined using seven and eight chemicals (mostly phenyl ureas, $r^2 = 0.96$) with a range of $\log K_{OW}$ from -0.7 to 4.3 (Briggs et al. 1982).
a : empirical correction factor for differences between lipids and octanol; for dimensional reasons, 'a' should be $\rho_{water}/\rho_{octanol}$, where $\rho_{octanol} = 0.822$, when sorption to lipids is the same as to octanol.

Equation 11.8 is often given as log / log regression of the form

$$\log(BCF - W) = b \log K_{OW} + c \tag{11.9}$$

where $c = 10^{L_P\, a}$

In the literature, the following regressions can be found:

$$\log(SXCF - 0.82) = 0.95 \log K_{OW} - 2.05 \tag{11.9a}$$

$SXCF = 0.82 + 0.0089K_{OW}^{0.95}$

where $SXCF$ = stem xylem concentration factor (Briggs et al. 1983). Mainly phenyl ureas, range $\log K_{OW}$ between -0.7 and 4.3, $n = 8$, $r^2 = 0.96$, macerated barley shoots.

$$\log(RCF - 0.82) = 0.77 \log K_{OW} - 1.52 \qquad (11.9b)$$

$$RCF = 0.82 + 0.03K_{OW}^{0.77}$$

where RCF = root concentration factor (Briggs et al. 1982, Mainly phenyl ureas, range $\log K_{OW}$ between -0.7 and 4.3, $n = 7$, $r^2 = 0.96$, mazerated barley roots.

$$\log(RCF - 0.85) = 0.557 \log K_{OW} - 1.34 \qquad (11.9c)$$

$$RCF = 0.85 + 0.046K_{OW}^{0.557}$$

Trapp and Pussemier (1991), 12 Carbamates, $1.16 < \log K_{OW} < 3.21$, $n = 12$, $r^2 = 0.92$, cut bean roots and stems.

$$\log(RCF - 0.85) = 0.75 \log K_{OW} - 1.96 \qquad (11.9d)$$

$$RCF = W_P + L_P K_{OW}^{0.75}$$

Trapp and Pussemier (1991), 12 Carbamates, $1.16 < \log K_{OW} < 3.21$, cut bean roots and stems; 0.85 is the water content and $L_P = 0.011$ is the lipid content (1.1%).

Estimation in the WATER model, bioconcentration factors for fish:

Veith found the following regression ($n = 84$; $r^2 = 0.82$):

$$\log BCF = 0.76 \log K_{OW} - 0.23 \qquad (11.10)$$

Species: blue gill sunfish, fathead minnow, mosquitofish, rainbow trout; mostly chlorinated hydrocarbons; $0.89 \leq \log K_{OW} \leq 6.9$.

By using a large number of literature values, Isnard and Lambert (1988) came to a similar relationship ($n = 107$, $r^2 = 0.82$; $0.98 \leq \log K_{OW} \leq 6.89$, several substance classes):

$$\log BCF = 0.80 \log K_{OW} - 0.52 \qquad (11.11)$$

and related to water solubility S (kg/m^3 = g/L; $2 \cdot 10^{0-7}$ g/L $\leq S \leq 36.308$ g/L, $r^2 = 0.75$):

$$\log BCF = 3.13 - 0.51 \log(1000 \cdot S) \qquad (11.12)$$

From **bioconcentration factors** BCF (unit mass substance per mass biota to mass substance per mass water), **partition coefficients** K (unit mass substance per volume biota to mass substance per volume water) can be calculated with the densities:

$$K_{BW} = BCF \, \rho_B / \rho_W$$

For fish with a density similar to water there is no difference.

11.7
Molecular Diffusion Coefficients

The diffusion coefficient in liquid or gaseous phase is often used. For the purpose of environmental exposure prediction, the estimation from the ratio of molar masses is sufficient (Eq. 3.2)

$$D_i/D_j = \sqrt{M_j}/\sqrt{M_i} \tag{11.13}$$

M is the molar mass (g/mol), i and j are indices for two chemicals, one of them a reference chemical with a known diffusion coefficient. The diffusion coefficients of oxygen in aqueous solution, M (O_2) = 32 g/mol, and of water vapor in air, M (H_2O) = 18 g/mol, serve as reference values.

$$D_G(H_2O) \quad = \quad 2.57 \cdot 10^{-5}\,m^2/s = 2.22\,m^2/d$$
$$D_W(O_2) \quad = \quad 2.0 \cdot 10^{-9}\,m^2/s = 1.728 \cdot 10^{-4}\,m^2/d$$

Exercises on chapter eleven

Uncertainty analysis
Brodsky (1986) determined water solubilities and $\log K_{OW}$ of 200 substances using several methods. For benzene, he found:

$$\log K_{OW} \quad = \quad 2.1 \quad \text{(literature value)}$$
$$\log K_{OW} \quad = \quad 2.45 \quad \text{(method 1)}$$
$$\log K_{OW} \quad = \quad 2.28 \quad \text{(method 2)}$$
$$\log K_{OW} \quad = \quad 2.39 \quad \text{(method 3)}$$
$$\log K_{OW} \quad = \quad 2.26 \quad \text{(method 4)}$$

water solubility S (kg/m^3)

$$S \quad = \quad 1.65 \quad \text{(literature value)}$$
$$S \quad = \quad 0.95 \quad \text{(method 1)}$$
$$S \quad = \quad 1.58 \quad \text{(method 2)}$$
$$S \quad = \quad 1.15 \quad \text{(method 3)}$$
$$S \quad = \quad 1.65 \quad \text{(method 4)}$$

Estimate the K_{OC} of benzene with the equations of Karickhoff (Eq. 11.3), Schwarzenbach and Westall (Eq. 11.4), Kenaga and Goring (Eq. 11.5) and Chiou (Eq. 11.6).

11.1 What is the maximal range of the resulting values of K_{OC}?

11.2 What is the velocity u_c of benzene in a sandy soil, with organic carbon OC between 1% and 3%, water content θ 10% to 30%, total porosity ε 30% to 50%, bulk soil density ρ_B 1.2 to 1.4 g/cm^3, flow velocity of the water between 0.2 and 2 cm/day and a K_{AW} of between 0.12 and 0.5?

a) Give the average of u_c (always use average data values)
b) Give the maximal range of u_c?
c) Which parameter has the greatest influence on u_c (see Chapter Seven for the equations)?

11.3 Program the equations so that you can vary the parameters randomly (uniform distribution) between the ranges given.

Monte-Carlo-Analysis: Use seven times 500 random numbers and vary all parameters simultaneously 500 times in the given ranges. Plot the resulting histogram of u_c. What is the distribution of u_c?

11.4 What is the probability (under the above conditions) for $u_c > 1$ m/a?

11.8
Substance Database

Data from Rippen (1996).

name: usually the trivial name, and not the IUPAC nomenclature

production t/a(year): worldwide production, in parenthesis: year
In many cases, environmental chemicals are not directly produced but are by-products of human activities, e.g. dioxins and PAH. In particular, gasoline by-products are often not produced, but imported.

M: molar mass in g/mol
$\log K_{OW}$: partition coefficient octanol-water
K_{AW}: partition coefficient air-water, dimensionless Henry's law constant, at 20°C

$t_{1/2}$ soil: approximate degradation half-life in soil, unit days; depends largely on environmental conditions; does not include volatilization and leaching. Often not of the 1st order (adaption time gives a lag phase). Qualitative data: short $t_{1/2} < 1$ week; long: some months; very long: no significant degradation.

λ_{OH} air: 2nd-order rate constant for degradation by [OH]-radicals, unit 10^{-12} cm^3/(s molecules). The mean concentration of [OH·] in air (temperate zone of the North) is approximately $5 \cdot 10^5$ molecules/cm^3. The pseudo first-order rate constant results from the multiplication $\lambda = \lambda_{OH} \cdot [OH·]$; e.g. ace-naphthen $\lambda = 79 \cdot 10^{-12}$ cm^3/(s molecules) $\cdot 5 \cdot 10^5$ molecules/cm^3 = $3.95 \cdot 10^{-5}$ s^{-1}. In some cases, half-life or qualitative data is given.

LC_{50} 96h: lethal concentration in water for fish species in mg/L (96h, if not given differently, usually static test) of S = trout *Salmo gairdneri*; F = fathead minnow *Pimephales promelas*; O = rainbow trout *Oncorhynchus mykiss*; L = *Leuciscus sp*; D = *Daphnia sp.*.

C = cancerogenous, M = mutagenous, ? = suspect.

Name	Sum formula	production t/a	(year)	M g/mol	log K_{OW} (–)	K_{AW} (–)	$t_{1/2}$ soil days	λ_{OH} air $\cdot 10^{-12}$	LC_{50} 96h species	mg/L	Cancerogenous Mutagenous
acenapthen	$C_{12}H_{10}$	2 500	(85)	154.21	4.20	0.0052	0.3–96	79	F	1.73	M
aldrin	$C_{12}H_8Cl_6$	nd	(nd)	364.91	5.52	0.021	180	nd	F	0.0083	
aniline	C_6H_7N	600 000	(nd)	93.15	0.95	$6 \cdot 10^{-5}$	2	116	S	20	C?
atrazine	$C_8H_{14}ClN_5$	90 000	(85)	215.69	2.58	$7.9 \cdot 10^{-8}$	71	14	F	15	
benzene	C_6H_6	17 500 000	(85)	78.12	2.12	0.23	5	1.1	F	33	M,C
benzo(a)pyrene	$C_{10}H_{12}$	none[a]		252.3	6.04	$1.39 \cdot 10^{-5}$	2–700	72.6	D	0.005	M,C
α-BHC	$C_6H_6Cl_6$	100 000	(83)	290.83	3.77	$9.6 \cdot 10^{-4}$	130	0.38	D	0.8	
β-BHC	$C_6H_6Cl_6$	15 000	(83)	290.83	3.85	$1.6 \cdot 10^{-5}$	2920	0.14	?	2.0	
γ-BHC lindane	$C_6H_6Cl_6$	38 000	(83)	290.83	3.66	$8.3 \cdot 10^{-5}$	260	0.24	F	0.059	M?C?
chlorobenzene	C_6H_5Cl	480 000	(85)	112.56	2.78	0.14	nd	0.82	F	30	
chloroform	$CHCl_3$	250 000	(80)	119.38	1.95	0.12	< 27	0.17	O	1.24–67	C
2,4-D	$C_6H_6Cl_2O_3$	50 000	(nd)	221.04	1.57	$< 1.6 \cdot 10^{-9}$	4–29	0.17	S	70	C,M?
p,p'-DDD	$C_{14}H_{10}Cl_4$	metabolite of DDT		320.05	6.02	< 0.001	> 3650	10	D	0.0082	C,M
DDE	$C_{14}H_8Cl_4$	metabolite of DDT		318.03	5.76	0.05	> 7300	40	O	0.032	C,M
DDT	$C_{14}H_9Cl_5$	5 000	(96)	354.49	6.20	0.0016	3650	10	F	0.01–0.045	M?,C?
DEHP	$C_{24}H_{38}O_4$	4 000 000	(nd)	390.56	7.48	$3 \cdot 10^{-4}$	8–98	28	F	> 770	?
dibutylphthalate	$C_{16}H_{22}O_4$	230 000	(79)	278.34	4.61	$2.6 \cdot 10^{-5}$	25	8.7	F	0.85–3.0	
1,2-diClbenzene	$C_6H_4Cl_2$	80 000	(nd)	147.01	3.40	0.07	100	0.4	F	57	
1,4-diClbenzene	$C_6H_4Cl_2$	80 000	(nd)	147.01	3.45	0.09	nd	0.44	F	30	
1,2-diClethane	$C_2H_4Cl_2$	33 600 000	(96)	98.96	1.46	0.053	long	0.25	F	118	C,M
dichloromethane	CH_2Cl_2	600 000	(96)	84.93	1.31	0.087	long	0.13	F	310	M,C?
dieldrine	$C_{12}H_8Cl_6O$	nd	(nd)	380.91	5.14	0.0044	870	nd	F	0.023–16.2	
ethanol	C_2H_6O	2 000 000	(nd)	46.07	–0.30	$2.1 \cdot 10^{-4}$	short	3.1	F	14 900	
ethene	C_2H_4	37 000 000	(85)	28.05	1.13	8.7	short	8.2	phytotoxic		
ethylbenzene	C_8H_{10}	10 000 000	(95)	106.16	3.20	0.24	nd	7.5	F	48	
freon 11	CCl_3F	310 100	(81)	137.38	2.53	7.2	> 40 a	< 0.0001	nd		
freon 12	CCl_2F_2	449 000	(81)	120.91	2.16	10.1	> 40 a	< 0.001	?	> 1000	

nd: no data
a) polycyclic aromatic hydrocarbon (PAH), product of firing processes

Name	Sum formula	production t/a	(year)	M g/mol	log K_{OW} (-)	K_{AW} (-)	$t_{1/2}$ soil days	λ_{OH} air $\cdot 10^{-12}$	LC_{50} 96h species	mg/L	Cancerogenous Mutagenous
hexaClbenzene	C_6Cl_6	5 000	(93)	284.79	5.80	0.028	730	< 0.6	F	22	C
LAS	$C_{18}H_{29}NaO_3S$	1 800 000	(87)	348.48	1.96	small	1.5–80	nd	S	1–4	
methanol	CH_4O	13 300 000	(86)	32.04	−0.71	0.000185	1–2	0.96	F	28 000	
nitrobenzene	$C_6H_5NO_2$	1 000 000	(nd)	123.11	1.85	0.0029	medium	0.17	F	117	M
OCDD	$C_{12}Cl_8O_2$	none	(96)	459.72	8.20	0.144?	very long	7.0	nd		
OCDF	$C_{12}Cl_8O$	none	(96)	443.72	8.54	nd	very long	nd	nd		
pentaClphenol	C_6HCl_5O	150 000	(83)	266.34	5.24	$2.9 \cdot 10^{-5}$	38	4.0	F	0.1–0.3	M?C?
phenol	C_6H_6O	4 000 000	(94)	94.11	1.44	$2.2 \cdot 10^{-5}$	1–30	28	F	33	
styrol	C_8H_8	10 070 000	(86)	104.14	3.07	0.09	70–700	55	F	29–59	
2,4,5-T	$C_8H_5Cl_3O_3$	3 000	(nd)	255.49	2.99	$2.7 \cdot 10^{-10}$	6.6–31	14	S	0.98	M,C?
2,3,7,8-TCDD	$C_{12}H_4Cl_4O_2$	none	(96)	321.97	6.76	0.0015	> 12 a	0.5–9.0	F	< 1.7e-6	C
2,3,7,8-TCDF	$C_{12}H_4Cl_4O$	none	(96)	305.96	6.31	0.0043	very long	2.3	S, LC_0	< 3.9e-6	?
tetraClethene	C_2Cl_4	750 000	(82)	165.83	2.87	0.54	very long	0.138	F	21	C?
tetraClmethane	CCl_4	1 000 000	(83)	153.82	2.77	0.93	long	< 0.0001	F	43	C
toluene	C_7H_8	30 000 000	(79)	92.15	2.66	0.23	< 7	6.1	F	12.6–42	C?
trichloroethene	C_2HCl_3	600 000	(80)	131.39	3.03	0.35	long	1.98	F	41–95	M,C
2,4,6-TNT	$C_7H_5N_3O_6$	nd	(96)	227.13	1.98	$1.1 \cdot 10^{-6}$	< 4–> 228	0.146	F	2.6–3.7	M,C
vinyl chloride	C_2H_3Cl	16 700 000	(85)	62.50	1.27	0.9	nd	5.8	L	1200	M,C
m-xylene	C_8H_{10}	> 3 340 000	(82)	106.17	3.18	0.22	5–60	23	S	8.4	

nd: no data

Solutions of the Exercises

Hint: Nobody's perfect. No guarantee is given for the results!

Exercises on Chapter Two

1) Degradation chain

$$\text{Tri} \xrightarrow{\lambda = 0.01d^{-1}} \text{Di} \xrightarrow{\lambda = 0.11d^{-1}} \text{VC} \xrightarrow{\lambda = 0.002d^{-1}} \text{Ethene}$$

a) concentration of tri after one year, Eq. 3.26 or 2.7:

$$C_t = C_0 \exp(-\lambda t) = 2.6 \, \text{mg/L}$$

b) concentration of vinyl chloride after one year

We need the analytical solution of

$$\frac{dC_2}{dt} = \lambda_1 C_1 - \lambda_2 C_2$$

analogous to example 2.3, Eq. 2.22;

$$C_2(t) = \frac{\lambda_1 C_1(0)}{\lambda_2 - \lambda_1} \left[\exp(-\lambda_1 t) - \exp(-\lambda_2 t)\right] = 56.75 \, \text{mg/L}$$

An interesting alternative to calculate this exercise by hand is the use of CemoS-CHAIN (chapter 10). When the transfer rates from one trophic level to the next are interpreted as the degradation rates, this exercise may be calculated nicely. We have prepared this in the file EX2.MIF. Import this file, run CHAIN (press F9), and you will get the result. The 'producer' is then trichloroethene, consumer 1 is vinyl chloride, and consumer 2 is ethene. Investigate the effect of degradation rates. It is interesting to see that trichloroethene itself vanishes relatively quickly, but that it takes several years until the metabolites have gone.

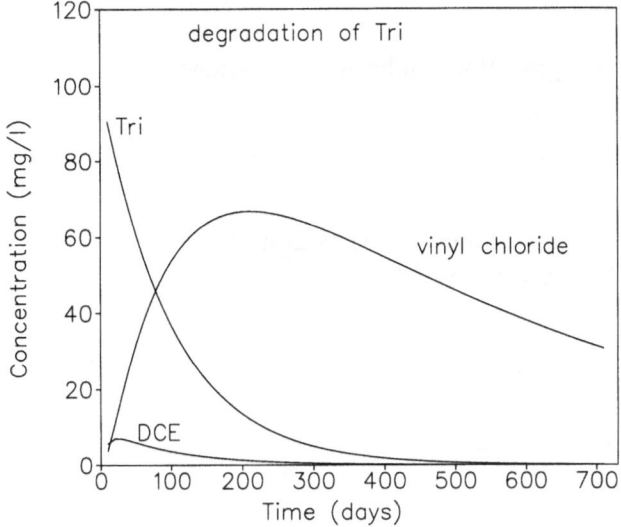

Figure: Degradation chain Tri to vinyl chloride

2) Ozone indoors

General equation:

$$dm/dt = \text{inflow} - \text{degradation} - \text{outflow} = Q\,C_{in} - k\,C\,V - Q\,C$$
$$dC/dt = dm/dt/V \;=\; C_{in}\,Q/V - kC - C\,Q/V$$
$$\;=\; C_{in}\,Q/V - (k + Q/V)\,C = b + aC$$

$$C(t) = C_0 \exp(-at) + b/a[1 - \exp(-at)]$$

with $C_0 = 0$
$k = \ln 2/5 = 0.1386 \text{ min}^{-1}$
$Q/V = 0.04 \text{ min}^{-1}$
$C_{in} = 300 \text{ μg/m}^3,\; b = 12 \text{ μg m}^{-3}\text{ min}^{-1}$
$a = Q/V + k = 0.1786 \text{ min}^{-1}$

2 a)
steady-state: $b/a = 67.2 \text{ μg m}^{-3}$
after 5 min:
$C(5\,\text{min}) = b/a\left[1 - \exp(-at)\right] = 39.7 \text{ μg m}^{-3}$
$C(60\,\text{min}) = 67.2 \text{ μg m}^{-3}$

2 b) theoretically never!

2c) deviation from steady-state below 5%:
$t(0.95)$ is when $\exp(-at) = 0.05$
$t(0.95) = 16.8 \text{ min}$

2d) dosage

$$dm/dt = \text{inflow} - \text{degradation} - \text{inhalation} - \text{outflow} =$$

$$Q\,C_{in} - k\,C\,V - Q\,C - I\,C$$

$I = 1/60\,\text{m}^3/\text{min}$

$a = I + Q/V + k = 0.1952\,\text{min}^{-1}$

$C = b/a = 12\,\mu g\,\text{m}^{-3}\,\text{min}^{-1}\,/\,0.1952\,\text{min}^{-1} = 61.48\,\mu g\,\text{m}^{-3}$

$\text{Dosage} = 61.48\,\mu g\,\text{m}^{-3}\cdot 1\,\text{m}^3/\text{h} = 61.48\,\mu g/\text{h}$

3) Aquarium

a) mass balance

$$\begin{aligned} dm_w/dt &= -k_1 m_w + k_2 m_f \\ dm_f/dt &= k_1 m_w - k_2 m_f \end{aligned}$$

3b) Analytical solution

This is not easy: first find the two eigenvalues, then the eigenvectors, and use the general solution for linear differential equation systems. Then find constants c1, c2 from initial conditions. Here we assume that initially, all mass is found in water and is $m_w(0)$. The result is:

$$m_w(t) = \frac{k_2 m_w(0)}{k_1 + k_2} + \frac{k_1 m_w(0)}{k_1 + k_2}\exp\left(-k_1 t - k_2 t\right)$$

$$m_f(t) = \frac{k_1 m_w(0)}{k_1 + k_2}\left[1 - \exp\left(-k_1 t - k_2 t\right)\right]$$

Concentrations: divide by volumes.

Exercises on chapter three

3.1 *Alarm!* Solution with Eq. 3.21

$$C(x,t) = \frac{m/a}{(4\pi D\,t)^{1/2}}\exp-\frac{(x - u\,t)^2}{4D\,t}$$

Approximate peak concentration is when $x = u\,t$; this simplifies the equation to

$$C_{max}(x) = \frac{m/A}{(4\pi D\,x/u)^{1/2}}$$

$$C_{max}(350\,000\,\text{m}) = 21.5\,\text{mg/L}$$

Note: The exact time of the peak concentration is found where $f'(x)$ is zero. It turns out that it is slightly before $u\,t = x$. The maximum concentration is approximately 1% above the calculated value.

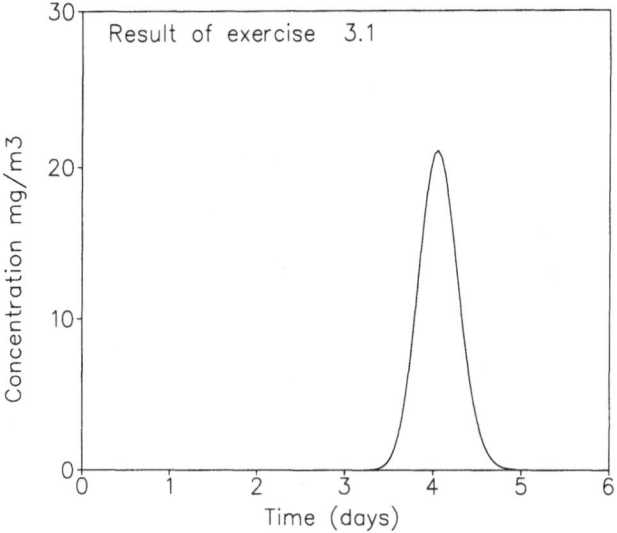

Figure: Plot of the result of exercise 3.1

3.2 Later ...

Concentrations are additive. The second pollution wave arrives 12 hours or 43 200 s later. The first peak in Düsseldorf is at $t = x/u = 350\,000/1$ s. So we use for the equation of the second peak the time $t = 350\,000$ s $- 43\,200$ s $= 306\,800$ s

$$C_2(x, t) = \frac{m/A}{(4\pi D\, t)^{1/2}} \exp - \left[\frac{(x - u\, t)^2}{4D\, t}\right]$$

$$= 11.39\,\mathrm{mg/m^3} \cdot \exp(-3.04) = 0.54\,\mathrm{mg/m^3}$$

Added to $C_{\max}$ of the first peak gives:

$$C = 22.4\,\mathrm{mg/m^3} = 22.4\,\mu\mathrm{g/L}$$

The arrival of the first peak is not necessarily the maximum concentration. The best way for the solution of the exercise is to program the equation and plot the result.

3.3 Yes, when the values for depth are interpreted as river length and time is the flow time. The equation is identical to 3.21 and could be used for the calculation of riverine problems, too. The program requires data for K_d and K_{AW} for the calculation of fractions in soil matrix, gas and water. Of course, those results are meaningless in this case. To test this, you may also import the file EX3_3.MIF, which contains this exercise.

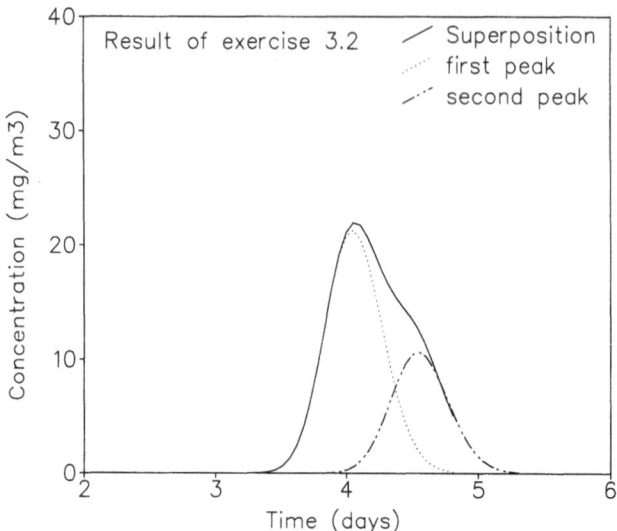

Figure: Plot of the result of exercise 3.2

Exercises on chapter four

4.1 Eq. 4.8 and 4.11a (or b)

$$K_d = OC\,K_{OC}$$

$$\log K_{OC} = 0.72\log K_{OW} + 0.49 \tag{4.11b}$$

$$K_{OC} = 43\,371, \quad K_d = 4\,337.1$$

$$K_{SW} = C_S/C_W = K_d\rho_S/\rho_W + \theta = 8675$$

$$C_S = C_W K_{SW} = 8.675 \cdot 10^{-3}\,\text{kg/m}^3$$
$$= 4.3\,\text{ppm} \quad (\text{mg/kg})$$

4.2 Eq. 4.8 and 4.11a (or b)

$$K_d \text{ of neutral species} = K_{OC}OC = 0.411 \cdot 10^{1,57} \cdot 0.1 = 1.53$$

fraction of neutral species, Eq. 4.12:

$$\Phi = \frac{1}{1 + 10^{a(\text{pH}-\text{pKa})}}$$

$$\text{pH} = 3: \quad \Phi = 0.35$$
$$\text{pH} = 5: \quad \Phi = 0.0053$$
$$\text{pH} = 7: \quad \Phi = 5.3 \cdot 10^{-5}$$

$$K_d(\text{pH 3}) \;=\; 0.53$$
$$K_d(\text{pH 5}) \;=\; 0.008$$
$$K_d(\text{pH 7}) \;=\; 8.1 \cdot 10^{-5}$$

The $K_d(\text{pH})$ calculated for high pH-values does not necessarily exclude any sorption, except those by lipophilic interactions.

4.3 Solve Example 4.1 with the Euler method.

1) Replace dt in the differential equation by Δt, and dC_1 by ΔC_1

$$\Delta C_1 = A_{12}g(C_2 - C_1/K_{12})/V_1\Delta t$$

2) Calculate the first step, e.g. with $\Delta t = 1000\,\text{s}$.
$\Delta C_1 = 1\,\text{m}^2 \cdot 10^{-3}\,\text{m/s} \cdot [0.1\,\text{mg/m}^3 - (0\,\text{mg/m}^3/10000)]/0.1\,\text{m}^3 \cdot 1000\,\text{s} = 1\,\text{mg/m}^3$

3) Then C_1 after $1000\,\text{s}$ is:
$C_1(1000\,\text{s}) = C_1(0) + \Delta C_1 = 1\,\text{mg/m}^3$.

4) Repeat the steps 2 to 4, but now with $C_1 = 1\,\text{mg/m}^3$ and so on (or use an adequate program).

The selection of the time step is critical!

If Δt is too long $\rightarrow$ $(C_2 - C_1/K_{12})$ is wrong. The first ΔC_1 is too large. In the next step, C_1 is too large, and now ΔC_1 is too small. The solution oscillates.
   If Δt is too small $\rightarrow$ enormous computation effort, rounding errors.

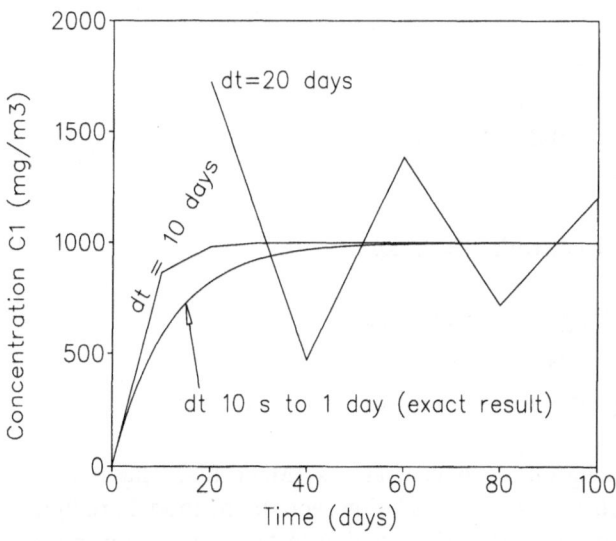

Figure: Numerical solution of example 4.1.

Solution of the problem: When you found a plausible solution, vary the time step. If the result changes, the numerical solution is most probably wrong.

The solution and the result with too large time steps are shown in the above Figure.

Exercises on chapter five

5.1

a) equilibrium partition coefficient between air and soil

$$K_{AB} = K_{AW}/K_{BW} \qquad \text{Eq. 4.13}$$

K_{AW} is 0.0015; plus Eq. 4.7

$$
\begin{aligned}
K_{BW} &= K_d \rho_B/\rho_W + \theta + (\varepsilon - \theta)K_{AW} \\
K_{BW} &= 0.411 \cdot 10^{6.76} \cdot 0.02 \cdot 1.5/1 + 0.3 + (0.5 - 0.3) \cdot 0.0015 = 7.1 \cdot 10^4 \\
K_{AB} &= 0.0015/7.1 \cdot 10^4 = 2.1 \cdot 10^{-8} \\
K_{BA} &= 1/K_{AB} = 4.7 \cdot 10^7 \quad \text{rather high, isn't it?}
\end{aligned}
$$

b) level 2 model, Eq. 5.4

$$C_1 = I/(\lambda_1 V_1 + \lambda_2 K_{21} V_2)$$

$1 = $ air, $2 = $ soil

$$
\begin{aligned}
V_1 &= 1.5 \cdot 10^{15} \, \text{m}^3; \\
V_2 &= 3.75 \cdot 10^{10} \, \text{m}^3 \\
\lambda_1 &= \ln 2/32 \, \text{d} = 0.021 \, \text{d}^{-1} = 7.9 \, \text{a}^{-1}; \\
\lambda_2 &= \ln 2/160 \, \text{a} = 0.0043 \, \text{a}^{-1}
\end{aligned}
$$

Now the equation can be solved, with $I = 0.1$ kg/a.

$$
\begin{aligned}
C_1 &= 5.15 \cdot 10^{-18} \, \text{kg/m}^3 = 5.15 \, \text{fg/m}^3 \\
C_2 &= C_1 \cdot K_{BA} = 0.242 \cdot 10^{-9} \, \text{kg/m}^3 \\
&= 161 \, \text{pg/kg (dry wt.)} \quad (1 \, \text{m}^3 \, \text{dry soil} = 1500 \, \text{kg})
\end{aligned}
$$

Compared to measured background concentrations (McLachlan, Bayreuth): Soil 70 pg/kg (dry weight), air 3.6 fg/m^3 (2.9 fg/m^3 gaseous). The model predicts the same order of magnitude. Uncertainties are due to the degradation rates and the amount emitted.

c) amounts in air and soil, steady-state

$$m = V C$$

soil: $m = 0.242 \cdot 10^{-9} \, \text{kg/m}^3 \cdot 3.75 \cdot 10^{10} \, \text{m}^3 = 9.1 \, \text{kg}$
air: $m = 5.15 \cdot 10^{-18} \, \text{kg/m}^3 \cdot 1.5 \cdot 10^{15} \, \text{m}^3 = 0.008 \, \text{kg}$

In steady-state, the majority of the dioxin is in the soil – about 90 times the annual emission! The degradation, however, occurs in air. So the next question:

d) half-life in the system, Eq. 5.5

$$\lambda_S = I/m$$

half-life $t_{1/2} = \ln 2 / \lambda_S = 63 \, \text{years}$

It is assumed here that free and immediate exchange between soil and air occurs. Indeed, dioxin is very strongly bound to the soil matrix. So it seems likely that the concentration in air decreases faster than assumed in this model. The result of the calculation is also shown in example 5.1.

You can also use the File EX5_1.MIF. Import into CemoS, run Level 2, and export the file. Type the command UREPORTF EX5_1.MIF EX5_1.TXT, and hopefully you get something like this:

```
Substance data for substance: 2,3,7,8-Tetrachlordibenzo-p-dioxin Descrip-
tion:
The famous ultra-toxic 'Seveso-dioxin'
Values taken from Rippen 1993;
Greatings from Trapp 1996
  CAS                                              1746-01-6   [-]
  Sum formular                                     C12H4Cl4O2  [-]
  Molecular weight                                   321.975   [g/mol]
  > estimated Trapp 9/93
  because of Estimation from sumformular
  KAW                                                 0.0015   [-]
  > calculated; Rippen 1993
  KOC                                            2.36506e+06   [cm3/g]
  > Range 910000 to 38 Mio. measured; Rippen 1993
  because of Estimation by Karickhoff
  depends on:
     log KOW
  log KOW                                               6.76   [-]
  > Rippen 1993
  Degradation rate in soil                       1.13324e-05   [1/d]
  > t12 = 60000 a
  > Exercise 5.1
  Degradation rate in air                             0.0216   [1/d]
  > Atkinson, calculated.
  > from Rippen 1993
  Vapor pressure                                     9.4e-05   [Pa]
  > Rordorf; subcooled liquid.
```

Model run for model: Level2 at: 07-17-1996 12:18:49

Size of soil area	2.5e+11	[m²]
Concentration of pollutant thruflux via air	0	[kg/m³]
Compartments: Air, Soil		

Density of the soil	1500	[kg/m³]
Soil depth	0.15	[m]
Height of the compartment air	6000	[m]
Initial mass in system	0	[kg]
Partition coefficient soil/water	70952	[-]
because of Estimation from K_{AW} and model parameter		
Partition coefficient soil matrix/water (K_d)	47301.2	[cm³/g]
because of Estimation from OC and K_{OC}		
Content of organic carbon particles in soil	0.02	[kg/kg]
Air volume flow	0	[m³/s]
Fraction of pore volume of dry soil	0.5	[m³/m³]
Pore water content	0.3	[m³/m³]
Time step of simulation	10	[a]
Time of simulation	1000	[a]
Release rate of substance	0.1	[kg/a]
PH value of soil water	6	[m]

Results:

Halflife	64.2146	[a]
Time to reach 95% steady state	277.531	[a]
Systemlossrate	0.0107942	[1/a]

Concentration at steady state [kg/m³]:

Air	5.21838e-18
Soil	2.46837e-10
Sum	2.46837e-10

Mass distribution at steady state [kg]:

Air	0.00782758
Soil	9.25638 (99.915507%)
Sum	9.2642 (100.000000%)

Table: Concentration profile (3 Columns):

2 := Concentration in air
3 := Concentration in soil

Time [a]	*2* [kg/m³]	*3* [kg/m³]
0	0	0
10	5.33949e-19	2.52565e-11
20	1.01326e-18	4.79287e-11
30	1.44353e-18	6.82811e-11

40	1.82978e-18	8.6551e-11
50	2.1765e-18	1.02952e-10
60	2.48775e-18	1.17674e-10
70	2.76715e-18	1.3089e-10
80	3.01796e-18	1.42754e-10
90	3.24311e-18	1.53404e-10
100	3.44522e-18	1.62964e-10
200	4.61588e-18	2.18337e-10
300	5.01366e-18	2.37153e-10
400	5.14882e-18	2.43546e-10
500	5.19475e-18	2.45719e-10
600	5.21035e-18	2.46457e-10
1000	5.21828e-18	2.46832e-10

5.2 Lindane in an old house.

a) equilibrium conditions: $\rightarrow C_A = K_{Awood}C_{wood}$
$K_{Awood} = 10^{-4}/1000 = 10^{-7}$

Explanation: It is much easier to measure the partition coefficient between water and wood than between air and wood which is rather high. The partition coefficient air to wood is then derived from K_{AW}

$$C_{wood} = 100\,\text{mg/kg} = 50\,\text{mg/l} = 0.5\,\text{g/m}^3$$

$$C_A = K_{Awood} \cdot C_{wood} = 0.5\,\text{g/m}^3\,10^{-7} = 0.05\,\mu\text{g/m}^3$$

b) Inhalation $1\,\text{m}^3/\text{h} \rightarrow$ uptake of $0.05\,\mu\text{g/h}$

Semivolatile chemicals such as lindane will also sorb to dust and textiles, they are taken up by direct contact with the skin, while they are applied to wood, and via textiles. Indeed, in a house where lindane was used for wood protection we found it in wood, dust, textiles and house plants in the ppm range (unpublished).

5.3 ... open the window
Calculation: Assumption of equilibrium; dynamical model level 2.

$$dm/dt = \text{Input} - \text{Output}$$

Input $= 0$;   Output $= Q\,C_A$

$$dm/dt = -Q\,C_A = dm_A/dt + dm_{wood}/dt \quad \text{(equilibrium)}$$

$$V_A dC_A/dt + V_{wood}dC_{wood}/dt = -Q\,C_A$$

with $C_{wood} = K_{woodAir}C_A$:

$$V_A dC_A/dt + K_{woodAir}V_{wood}dC_A/dt = -Q\,C_A$$

$$dC_A/dt = -C_A \, Q/(VA + K_{woodAir} V_{wood})$$

$$C_A(t) = C_A(0) \exp(-at)$$

$$a = \frac{Q}{V_A + K_{woodA} V_{wood}} = \frac{1 \, m^3 \, min^{-1}}{50 \, m^3 + 0.1 \cdot 10^7} \approx 10^{-6} \, min^{-1}$$

What time corresponds to a remaining mass of 1%?

$$\ln[C_A(t)/C_A(0)] = -at$$

$$t = 8.8 \, years$$

Even if lindane has a quick exchange with the room air, and this is likely because of the high surface area, only a very small fraction remains in the air due to the high partition coefficent. The opening of the window decreases the air concentration of toxic wood preservation chemicals, but it increases again after closing the window. This is different for more fugitive chemicals such as the spray insectizide dichlorvos, or formaldehyde. Those chemicals have initially much higher concentrations in room air, but they do not sorb as strongly and (hopefully) vanish faster.

5.4 CemoS Level 2

		concentration	mass balance
water	:	9.58E-14 kg/m^3	6.84E-4 kg
air	:	5.17E-15 kg/m^3	11.19 kg
soil	:	1.17E-10 kg/m^3	8.33 kg
sediment	:	1.94E-10 kg/m^3	0.024 kg
suspense.	:	5.83E-10 kg/m^3	1.39E-4 kg
fish	:	1.15E-9 kg/m^3	0.0041 kg
plant	:	5.20E-10 kg/m^3	0.186 kg
			19.7335325 kg

Comment: The volume of the air is highest in the standard scenario FRG. Although small concentrations of HCB are calculated, the majority of the chemical is in the air. Highest concentrations occur in biota.

Exercises on chapter six

6.1

a) Tetrachloroethene, $K_{AW} = 0.83 \gg 0.04$, only liquid resistance has to be considered.

$$k_l = 0.2351 u^{0.969} \, h^{-0.673} (32/M)^{0.5} = 0.0285 \, m/h$$

$$1/K_V = h\,[1/(k_l) + 1/(K_{AW} k_g)]$$

$K_{AW} k_g \gg k_l$:

$$K_V = \text{ca. } k_l/h = 0.0114\,\text{h}^{-1}$$

Tetrachloroethene volatilizes quickly, the half-life is 2.5 days.

b) Atrazine:

$K_{AW} < 4 \cdot 10^{-6}$: Volatilization small.

$1/K_V = h[1/(k_l) + 1/(K_{AW} k_g)]$

$K_{AW} k_g \ll k_l$, gas film controlled, only k_g is of importance

$k_g = 11.37\,(v + u)\,(18/M)^{0.5} = 7.55\,\text{m/h} = 181.2\,\text{m/d}$

$K_V = \text{ca. } K_{AW} k_g/h = 5.8 \cdot 10^{-7}\text{d}^{-1}$

The half-life is more than 1000 years. The volatilization of atrazine from rivers into air can be neglected.

6.2 Sewage purification plant

a) A possible solution of this exercise is found in Chapter 6.6.

$$dm/dt = Q_{in} C_{in} - \lambda V\, C_W(1 + R\,K_d) - R\,Q_{in}\,K_d\,C_W - Q_A\,C_W\,K_{AW} - Q_{in}\,C_W$$

steady-state: $dm/dt = 0$

$$C_W = \frac{Q_{in} C_{in}}{\lambda V(1 + R\,K_d) + R\,Q_{in}\,K_d + Q_A\,K_{AW} + Q_{in}}$$

b) DEHP $C_W = 19.2 \cdot 10^{-6}\,\text{kg/m}^3 = 0.019\,\text{mg/l}$
c) Dichlorobenzene $C_W = 3.25\,\text{mg/m}^3 = 0.32\,\mu\text{g/l}$
d) Only if you are interested in waste water alone:

DEHP : $C_W/C_{in} = 1.9\%$ (cleaning $> 98\%$).

DCB : $C_W/C_{in} = 32\%$ (cleaning $> 66.67\%$)

The substances have not vanished. More than 95% of DEHP remain in sewage sludge, i.e. degradation $< 3\%$ in input, 33.5% of dichlorobenzene remain in sewage sludge, the same amount volatilizes, i.e. degradation $< 2\%$. Note that these results are not validated by experiments at real WWTPs.

6.3 WATER-mass balance trichloroethene:

volatilization	:	63.14%
sedimentation	:	0.0042%
degradation	:	0.117%
advection	:	36.73%
Sum		100.0%

WATER-mass balance atrazine:

volatilization	:	0.0025%
sedimentation	:	0.0167%
degradation	:	1.835%
advection	:	98.14%
Sum		100.0%

Trichloroethene degrades slowly, but it is eliminated from rivers by volatilization. Atrazine, too, is not readily degraded, but it has a low vapor pressure and a small K_{AW}, and volatilization is not a significant fate process. Almost all mass is advected downstream.

The files EX6_3A.MIF and EX6_3B.MIF contain the exercise.

Exercises on chapter seven

7.1 Basic equation $D_c = D_{W,a} f_w + D_{G,eff} f_g$
needed: f_w, f_g, $D_{W,a}$, $D_{G,eff}$

$$f_w = \frac{\theta}{K_{MW} + \theta + (\varepsilon - \theta)\Phi K_{AW}}$$

$$f_g = \frac{(\varepsilon - \theta)\Phi K_{AW}}{K_{MW} + \theta + (\varepsilon - \theta)\Phi K_{AW}}$$

Benzene:

$$K_{MW} = K_d \rho_B/\rho_W = K_{OC}\, OC\, \rho_B/\rho_W = 0.411 K_{OW}\, \Phi\, OC\, \rho_B/\rho_W$$

K_{MW} (Benzene) = 1.35 (no dissociation, $\Phi = 1$);

$$f_w = 0.177$$
$$f_g = 0.027$$

Phenol:

dissociation, $\Phi =?$;   Eq. 4.12

$\Phi = [HA]/\{[HA] + [A-]\} = [1 + 10^{(7-9,9)}]^{-1} = 0.9987$
(practically no dissociation at pH 7, $\Phi \approx 1$)

K_{MW} (Phenol) = 0.322 (estimation of K_{OC} with Karickhoff's equation);
$f_w = 0.48$; $f_g = 7.07 \cdot 10^{-6}$

order of magnitude of the diffusion:
in gases: approximately 1 m²/d, in water 10^{-4} m²/d
dispersion: $D_{Disp} = q\,L = 0.001$ m/s $\cdot$ 0.05 m $= 5 \cdot 10^{-5}$ m²/d

$$D = D_{W,a} f_w + D_{G,eff} f_g$$

and $D_{W,a} \ll D_{G,eff}$ gives:

Benzene diffuses much faster than phenol due to the higher fraction present in gas pores.

The flow velocity of the chemical can also be calculated by:

$$u_c = u_w f_w$$

$$f_w(\text{phenol}) = 0.48 > f_w(\text{benzene}) = 0.177$$

Phenol leaches faster than benzene.
Compare the calculation in CemoS, EX7_1A.MIF and EX7_1B.MIF.

7.2 Use the SOIL model

formaldehyde	$7.88 \cdot 10^{-4}$ m/d
phenol	$2.08 \cdot 10^{-4}$ m/d
benzene	$1.12 \cdot 10^{-4}$ m/d
trichloroethene	10^{-4} m/d
nitrobenzene	$6.48 \cdot 10^{-5}$ m/d
atrazine	$4.4 \cdot 10^{-5}$ m/d
2,4,6-trichlorophenol	$4 \cdot 10^{-5}$ m/d
pentachlorophenol	$2.3 \cdot 10^{-5}$ m/d
hexachlorobenzene	$2.83 \cdot 10^{-7}$ m/d
benzo[a]pyrene	$1.67 \cdot 10^{-7}$ m/d
DEHP	$1.83 \cdot 10^{-8}$ m/d
2,3,7,8-TCDD	$6.6 \cdot 10^{-9}$ m/d

Exercises on chapter eight

8.1 Calculate with the box model AIR

Concentration of benzene in the box
Dose in the box (kg/a)
Deposition in the box kg/(m^2 a)

You can use the MIF-File EX8_1.MIF. Import it, run AIR, then export it, use the command UREPORTF EX8_1.MIF EX8_1.TXT.

Substance data for substance: Benzene
    Trapp 7/96

Molecular weight	78.1136	[g/mol]
> estimated Trapp 9/93		
K_{AW}	0.23	[-]
> measured at 25 degree C; from Rippen 1993		
K_{OC}	89	[cm³/g]
> measured, from Rippen 1993		
$\log K_{OW}$	2.12	[-]
>from Rippen 1993		
Degradation rate in soil	0.021	[1/d]
> Half-life 5 to 60 days; measured		
> from Rippen 1993		
Degradation rate in air	0.0475	[1/d]
> $l(OH) = 1.1e^{-12}\,cm^3/(smol)$; from Rippen 1993.		
Vapor pressure	10100	[Pa]
> Rippen 1993		

Model run for model: Air   at: 07-17-1996   11:23:21

Particulate fraction	2.52475e-09	[-]
because of Estimation from vapor pressure		
Inhalation rate	20	[m³/d]
Release rate of substance	0.5123	[kg/d]
Precipitation rate	2.1	[mm/d]
Vertically averaged wind speed	0	[m/s]
Dry gas deposition velocity	0.00979867	[m/s]
because of Estimation from molar mass		
Dry particle deposition velocity	0.01	[m/s]
Length of box in direction of the wind	1000	[m]
Width of box	1000	[m]
Height of box	500	[m]

Results:

Concentration in box	5.88604e-10	[kg/m³]
Total deposition	0.000181887	[kg/m²a]
Inhalation dose	4.29681e-06	[kg/a]

Mass balance [kg/a]:

Dry particle-bound deposition	4.6865e-07	(0.000000%)
Deposition by particle washout	2.27816e-07	(0.000000%)
Dry gaseous deposition	181.885	(97.270208%)
Deposition by gas washout	0.00196159	(0.001049%)
Degradation	5.10246	(2.728742%)
Advection	0	(0.000000%)
Sum	186.989	(100.000000%)

If you prefer a calculation by 'hand', follow the following steps:

Basic equation 8.1

$$C_0 = \frac{I}{u\, y_0\, z_0 + v_{dep} x_0 y_0 + \lambda V}$$

needed: v_{dep}

four deposition velocities: dry and wet deposition, each gaseous and particle bound.

fraction of gaseous and particulate compound:

$$K_{gp} = p/(c\, A_T)$$

where

$$p = 10\,100\,\text{Pa}, \quad c = 17\,\text{Pa\,cm}, \quad A_T = 1.5 \cdot 10^{-6}\,\text{cm}^2/\text{cm}^3$$

$$f_p = c_{par}/(c_{gas} + c_{par}) = 1/(1 + K_{gp}) = 2.5 \cdot 10^{-9}$$

f_p, the particle bound fraction of benzene, can be neglected.

Dry gaseous deposition:
$$v_{d,gas} = v_{d,gas}(ref)[M(ref)/M]^{0.5} = 0.0098\,\text{m/s}$$

Wet gaseous deposition:
$$v_w = P/(K_{AW} \cdot 8.64 \cdot 10^7) = 10^{-7}\,\text{m/s}$$

It follows that dry gaseous deposition is the only relevant deposition process of benzene.

In total, $v_{dep} \approx v_{d,gas} = 0.0098\,\text{m/s}$.

Concentration in the box:

$$C_0 = \frac{I}{v_{dep} x_0\, y_0 + \lambda V} = 5.88 \cdot 10^{-10}\,\text{kg\,m}^{-3} = 588\,\text{ng\,m}^{-3}$$

Dosis in the box (kg/a):

$$D = i\, C_0 = 4.29\,\text{mg/a}$$

Total deposition in the box kg/(m²a)

$$\begin{aligned} d &= v_{dep} C_0 = 182 \cdot 10^{-6}\,\text{kg\,m}^{-2}\,\text{a}^{-1} \\ &= 182\,\text{kg\,a}^{-1} \text{ for the box.} \end{aligned}$$

8.2 Calculation of the same scenario with a Level 2 fugacity model. You can use the MIF-File EX8_2.MIF. Import it, run Level 2, then export it, use the command UREPORTF EX8_2.MIF EX8_2.TXT.

Substance data for substance: Benzene
Description:
    Trapp 7/96

Sum formular	C_6H_6	[-]
Molecular weight	78.1136	[g/mol]

> estimated Trapp 9/93

K_{AW}	0.23	[-]

> measured at 25 degree C; from Rippen 1993

K_{OC}	89	[cm^3/g]

> measured, from Rippen 1993

$\log K_{OW}$	2.12	[-]

> from Rippen 1993

Degradation rate in soil	0.021	[1/d]

> Half-life 5 to 60 days; measured
> from Rippen 1993

Degradation rate in air	0.0475	[1/d]

> $l(OH) = 1.1e - 12\,\text{cm}^3/(\text{s}\,\text{mol})$; from Rippen 1993.
> Trapp 9/93

Vapor pressure		[Pa]

Modelrun for model: Level 2   at: 07-17-1996   11:28:14

Size of soil area	1e+06	[m^2]
Concentration of pollutant thruflux via air	0	[kg/m^3]

Compartments: Air, Soil

Density of the soil	1500	[kg/m^3]
Soil depth	0.2	[m]
Height of the compartment air	500	[m]
Initial mass in system	0	[kg]
Partition coefficient soil/water	3.016	[-]

because of Estimation from K_{AW} and model parameter

Partition coefficient soil matrix/water (K_d)	1.78	[cm^3/g]

because of Estimation from OC and K_{OC}
    Content of organic carbon particles in soil

Content of organic carbon particles in soil	0.02	[kg/kg]
Air volume flow	0	[m^3/s]
Fraction of pore volume of dry soil	0.5	[m^3/m^3]
Pore water content	0.3	[m^3/m^3]
Time step of simulation	0.05	[a]
Time of simulation	1	[a]
Release rate of substance	187	[kg/a]
PH value of soil water	6	[m]

Results:

Halflife	0.0400964	[a]
Time to reach 95% steady state	0.173294	[a]
Systems loss rate	17.287	[1/a]

Concentration at steady state [kg/m^3]:

Air	2.15218e-08	
Soil	2.82217e-07	

Sum	3.03739e-07	

Masses at steady state [kg]:

Air	10.7609	(99.478215%)
Soil	0.0564433	(0.521785%)

Sum	10.8174	(100.000000%)

Now the concentration in air is $2.15 \cdot 10^{-8}\,\text{kg/m}^3$, which is 37 times higher!

8.3 The results of exercise 8.1 and 8.2 differ significantly. Why is the concentration in the AIR box-model so much lower?

The reason is: in the AIR box model, deposition is a sink. In reality, benzene might desorb; when a thermodynamical equilibrium between upper soil and atmosphere is reached, then no net deposition occurs. This situation is calculated by the Level 2 multi-media model (chapter Five). In this case, it is the more realistic model. AIR is appropriate for the prediction of deposition pathways.

And why is the concentration calculated with Level 2 higher than the one calculated in Chapter Five? Here, you have only 500 m atmospheric mixing height. This is a 12 times smaller volume as in Chapter Five, and the resulting concentration is approx. 12 times higher.

8.4 a) equilibrium concentration of benzene in butter

$$
\begin{aligned}
K_{\text{butter/air}} &= K_{OW}/K_{AW} = 547 \\
C_{\text{butter}} &= 0.05\,\text{mg/m}^3 \cdot 547 = 27.4\,\text{mg/m}^3 \\
m_{\text{butter}} &= 27.4\,\text{mg/m}^3 \cdot 0.3 \cdot 10^{-3}\,\text{m}^3 = 8.2\,\mu\text{g}
\end{aligned}
$$

b) $D = i \cdot C_{air} = 10\,\mu\text{g}$
c) $D = i \cdot C_{air} = 200\,\mu\text{g}$

The urban air concentration of benzene leads to the highest uptake. This increases the risk of cancer. Literature: LAI 1992 (see example of Chapter Five, too).

Exercises on chapter nine

9.1 Uptake from soil 9.2

a) Eq. 4.6
$$C_W = C_M/K_d = 5\,\text{mg/kg}\,/\,0.58\,\text{kg/l} = 8.62\,\text{mg/l}$$
b) Eq. 9.3–9.4ab:
(9.4a) $TSCF = 0.784\exp[-(\log K_{OW} - 1.78)^2/2.44] = 0.7824$
(9.4b) $TSCF = 0.7\exp[-(\log K_{OW} - 3.07)^2/2.78] = 0.41$
result of Eq. 9.4a is higher and is taken

$$m = Q_W\,CW\,TSCF\,t = 1\,\text{L/d} \cdot 8.62\,\text{mg/L} \cdot 0.7824 \cdot 50\,\text{d} = 337.21\,\text{mg}$$

9.2 Uptake from air 9.3

a) $\begin{aligned}
K_{LA} &= K_{LW}/K_{AW} \\
K_{OW} &= 10^{1.85} = 70.79 \\
K_{LW} &= (W_P + L_P\,a\,K_{OW}^{0.95})\rho_P/\rho_W \\
&= (0.8 + 0.02 \cdot 1.22 \cdot 70.79^{0.95})1/1 = 2.2 \\
K_{LA} &= 3600 \\
C_{\text{Lettuce}} &= C_A\,K_{LA} = 10.8\,\mu\text{g/kg}
\end{aligned}$

b) Eq. 9.12: $C_L(t) = b/a[1 - \exp(-at)]$

$\begin{aligned}
a &= \lambda_E + \lambda_w + A\,g/(K_{LA}\,V_L) \text{ and } b = A\,g\,C_A/V_L \\
a &= 1.75 \cdot 10^{-4}\,\text{s}^{-1} \\
b &= 1.8\,\mu\text{g m}^{-3}\,\text{s}^{-1}
\end{aligned}$

steady-state: $b/a = 10.3\,\mu\text{g/kg}$

$t = 1\,\text{hour} = 3600\,\text{s}$:

$$C_L(3600\,\text{s}) = 10.3\,\mu\text{g/kg} \cdot [1 - \exp(-at)] = 4.8\,\mu\text{g/m}^3$$

$t = 50$ days: steady-state

9.3 Competitive uptake from soil and air

$\begin{aligned}
a &= 1.75 \cdot 10^{-4}\,\text{s}^{-1}\ (\text{as before}) \\
b &= b_1\ (\text{uptake from soil}) + b_2\ (\text{uptake from air}) \\
&= Q_W\,TSCF\,C_W/V_L + A\,g\,C_A/V_L \\
&= 7.8 \cdot 10^{-5}\,\text{g}/(\text{m}^3\,\text{s}) + 0.18 \cdot 10^{-5}\,\text{g}/(\text{m}^3\,\text{s}) \\
&= 7.98 \cdot 10^{-5}\,\text{g}/(m^3\,s) \\
C_L &= b/a = 0.456\,\mu\text{g/kg}
\end{aligned}$

b_1 is much larger than b_2, nitrobenzene is taken up from soil where the concentration is much higher. After translocation, nitrobenzene volatilizes from leaves into air because $C_L/C_A \gg K_{LA}$, and net diffusion is always directed towards equilibrium.

Time to reach 95% steady-state

$$C_L(t)/C_L(\infty) = 0.95 \rightarrow 1 - \exp(-at) = 0.95$$

$$\rightarrow t = -\ln(0.05)/a = 4.25\,\text{h}$$

Exercises on chapter ten

Exercise 10.1

Results for $t = 60\,\text{d}$:
masses $m_{\text{grass}} = 81\,\mu\text{g}$
$m_{\text{Xarne}} = 11.66\,\mu\text{g}$
$m_{\text{milk}} = 7.27\,\mu\text{g}$ (at $t = 61\,\text{d}$: $m_{\text{milk}} = 7.47\,\mu\text{g}$)
concentrations $C_{\text{grass}} = 8.1\,\text{ng/kg}$
$C_{\text{Xarne}} = 23.3\,\text{ng/kg}$
$C_{\text{milk}} = (7.47423 - 7.2752\,\mu\text{g})/25\,\text{L} = 7.9\,\text{ng/l}$

You may also import the file EX10.MIF. However, only masses may be calculated.

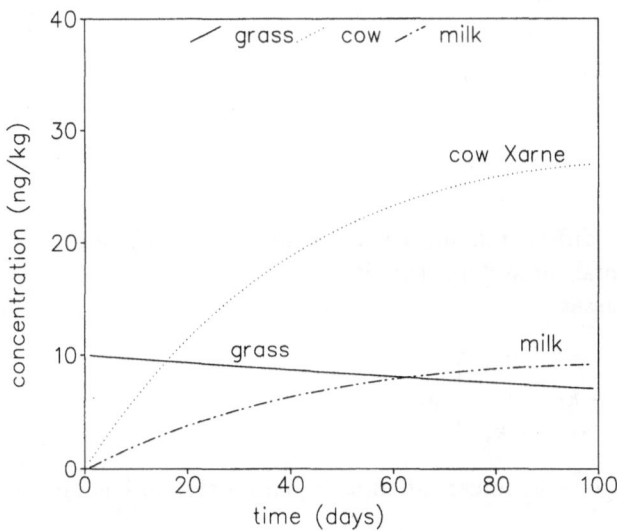

Figure: Solution of the exercise on Chapter Ten

If you import EX10.MIF, execute CHAIN (F9 key), export the MIF-File, and rewrite it with the awk-script UREPORTF, you get:

Model run for model: Chain   at: 07-17-1996   15:44:39

Degradation rate of substance in consumer 1	0	[kg]
Degradation rate of substance in consumer 2	0	[kg]
Degradation rate of substance in producer	0	[kg]
Single input into producer	1e-07	[kg]
Mass of consumer 1	500	[kg]
Mass of consumer 2	1500	[kg]
Mass of producer	10000	[kg]
Transfer coefficient consumer 1 → consumer 2	0.017	[1/d]
Transfer coefficient producer → consumer 1	0.0035	[1/d]
Duration of simulation	60	[d]
Number of output intervals	100	[-]

Results:

Masses [kg]:

Mass of substance in producer	8.10584e-08	(81.058425%)
Mass of substance in consumer 1	1.16664e-08	(11.666389%)
Mass of substance in consumer 2	7.27519e-09	(7.275186%)
Degradated mass	−2.01948e-27	(−0.000000%)
Sum	1e-07	(100.000000%)

10.2 Cascade, solution method

The general solution scheme for linear differential equation systems is applied to the cascade example. The system for $n = 3$ is:

$$dm_1(t)/dt \quad = \quad -k_1 m_1(t)$$
$$dm_2(t)/dt \quad = \quad +k_1 m_1(t) - k_2 m_2(t)$$
$$dm_3(t)/dt \quad = \quad +k_2 m_2(t) - k_3 m_3(t)$$

Written as matrix:

$$\dot{m}(t) = A\,m(t)$$

$\dot{m}(t)$ = vector of the differential quotients dm_i/dt (change of mass)
A = compartmental- or systems matrix
$m(t)$ = vector of masses

$$\dot{m}(t) = \left\{ \begin{array}{ccc} -k_1 & 0 & 0 \\ k_1 & -k_2 & 0 \\ 0 & k_2 & -k_3 \end{array} \right\} m(t)$$

The corresponding homogeneous linear diffential equation system has only a non-vanishing solution if

$$\det |A - \lambda E| = 0$$

$$\det \left\{ \begin{array}{ccc} -k_1 - \lambda & 0 & 0 \\ k_1 & -k2 - \lambda & 0 \\ 0 & k_2 & -k_3 - \lambda \end{array} \right\} = 0$$

For $n = 3$ the determinant can be found by the rule of *Sarrus*:

$$\begin{aligned} \det A \; &= \; a_{11}a_{22}a_{33} - a_{11}a_{23}a_{32} \\ &+ \; a_{12}a_{23}a_{31} - a_{12}a_{21}a_{33} \\ &+ \; a_{13}a_{21}a_{32} - a_{12}a_{22}a_{31} \end{aligned}$$

$$\det A \; = \; (-k_1 - \lambda)(-k_2 - \lambda)(-k_3 - \lambda)$$

For the condition $\det A = 0 = (-k_1 - \lambda)(-k_2 - \lambda)(-k_3 - \lambda)$ it follows

$$\lambda_1 = -k_1; \quad \lambda_2 = -k2; \quad \lambda_3 = -k_3$$

The eigenvectors **u** of A are the non-vanishing solutions of

$$(A - \lambda E)\mathbf{u} = 0 \quad \text{or} \quad \lambda E \mathbf{u} = A \mathbf{u}$$

With the first eigenvalue λ_1 for the eigenvector u_1

$$\begin{aligned} \lambda_1 u_{11} \; &= \; -k_1 u_{11} + 0 u_{12} + 0 u_{13} \\ \lambda_1 u_{12} \; &= \; +k_1 u_{11} + (-k_2) u_{12} + 0 u_{13} \\ \lambda_1 u_{13} \; &= \; 0 u_{11} + k_2 u_{12} + (-k_3) u_{13} \end{aligned}$$

since $\lambda_1 = -k_1$ it follows

u_{11} can be chosen arbitrarily, useful $= 1$

$$\begin{aligned} u_{12} \; &= \; k_1/(k_2 - k_1) \\ u_{13} \; &= \; (k_1 k_2)/[(k_2 - k_1)(k_3 - k_1)] \end{aligned}$$

Analogously with λ_2 and λ_3 for u_2 and u_3:

$$\begin{aligned} u_{21} \; &= \; 0; \quad u_{22} = 1; \quad u_{23} = k_2/(k_3 - k_2); \\ u_{31} \; &= \; 0; \quad u_{32} = 0; \quad u_{33} = 1; \end{aligned}$$

If these values are inserted into the general solution (Eq. 2.27)

$$\mathbf{m}(t) = c_1 \exp(\lambda_1 t)\mathbf{u}_1 + c_2 \exp(\lambda_2 t)\mathbf{u}_2 + \ldots + c_n \exp(\lambda_n t)\mathbf{u}_n$$

then

$$\begin{aligned} m_1(t) \; &= \; c_1 \exp(-k_1 t) \\ m_2(t) \; &= \; -\frac{c_1 k_1 \exp(-k_1 t)}{k_2 - k_1} + c_2 \exp(-k_2 t) \\ m_3(t) \; &= \; \frac{c_1 k_1 k_2 \exp(-k_1 t)}{(k_2 - k_1)(k_3 - k_1)} + \frac{c_2 k_2 \exp(-k_2 t)}{(k_3 - k_2)} + c_3 \exp(-k_3 t) \end{aligned}$$

The coefficients c_i follow from the initial conditions. With $m_1(t = 0) = m_0$, $m_2(t = 0) = 0$ and $m_3(t = 0) = 0$ we yield

$$
\begin{aligned}
m_1(0) &= c_1 \exp(-k_1 t); \quad \text{with } \exp(0) = 1 \text{ follows: } c_1 = m_0; \\
m_2(0) &= c_1 k_1/(k_2 - k_1) + c_2 = 0;
\end{aligned}
$$

It follows that

$$
\begin{aligned}
c_2 &= -m_0 k_1/(k_2 - k_1) \\
m_3(0) &= \frac{c_1 k_1 k_2}{(k_2 - k_1)(k_3 - k_1)} + \frac{c_2 k_2}{k_3 - k_2} + c_3 \\
c_3 &= \frac{m_0 k_1 k_2}{(k_1 - k_3)(k_2 - k_3)}
\end{aligned}
$$

Finally, we get

$$
\begin{aligned}
m_1(t) &= m_0 \exp(-k_1 t) \\
m_2(t) &= \frac{k_1 m_0}{k_2 - k_1}[\exp(-k_1 t) - \exp(-k_2 t)] \\
m_3(t) &= k_1 k_2 m_0 \left\{ \frac{\exp(-k_1 t)}{(k_2 - k_1)(k_3 - k_1)} + \frac{\exp(-k_2 t)}{(k_1 - k_2)(k_3 - k_2)} \right. \\
&\quad \left. + \frac{\exp(-k_3 t)}{(k_1 - k_3)(k_2 - k_3)} \right\}
\end{aligned}
$$

identical to equations 2.20 to 2.24.

Exercises on chapter eleven

Uncertainty analysis

11.1 maximum range 51.74 to 179.47

11.2 a) u_c average 0.055 cm/d
b) range of u_c is 0.0025 cm/d to 0.635 cm/d
c) flow velocity u

Can be seen directly from the equations plus the range; another way is to investigate the correlation matrix between result u_c and the input data of the Monte Carlo analysis:

	correlation
	to u_c
u	**0.612**
K_{OC}	−0.376
OC	−0.265
θ	0.410
ε	−0.010
ρ	−0.055
K_{AW}	−0.030

11.3 Histogram of u_c for 500 runs; each * represents 5 events.

Midpoint u_c [cm/d]	number	
0.00	43	*********
0.04	193	***
0.08	133	***************************
0.12	67	**************
0.16	34	*******
0.20	19	****
0.24	7	**
0.28	2	*
0.32	1	*
0.36	0	
0.40	1	*

The distribution seems to be log normal.

11.4 Probability $u_c > 1$ m/a (0.00274 m/s)

There are two ways of solutions:

a) statistical approach
   Take logarithm of Monte Carlo result u_c, test on normal distribution (yes, on a high significance level), mean of log u_c is −3.2185, standard deviation is 0.3273;
   Use the standard normal distributed variable z to predict probability:

   $$\log u_c(0.00274\,\text{m/d}) = -2.5622 = -3.2185 + z\,0.3273; \quad z = 2.005;$$

   Probability for $z \geq 2.00$
   $P(z \geq 2,00) < 0.023(2.3\%)$
   Some statistical error may occur, due to the small number of events (500).

b) The easier way. Use the law of big numbers (for $n \to \infty$, the relative frequency becomes the probability). Do the Monte Carlo analysis with very many runs, e.g. $n = 500\,000$; count the number of $u_c > 1\,\text{m/a} = 0.00274\,\text{m/d}$.

You can only use the method if you have a good random generator! In an example, we found from $500\,000$ runs 4947 times $u_c > 0.0027397\,\text{m/d}$, probability about 1%.

Random generator from P. L'Eculyer (1988), *Communications of the ACM* **31**, p. 742; cited in T. Otto (1994): Auf zufälligen Wegen zum Ziel. *Computertechnik* (1994) **5**, p. 264.

References

Abramowitz, M. and Stegun, I. (National Bureau of Standards, USA, 1972): Handbook of Mathematical Functions with Formulas, Graphs and Mathematical Tables. John Wiley and Sons, New York, 10th ed.

Ahlers, J., Diderich, R. Klaschka, U., Marschner, A. and Schwarz-Schulz, B. (1994): Environmental Risk Assessment of Existing Chemicals. *ESPR - Environ. Sci. & Pollut. Res.* 1(2), 117–123.

Aho, A. V., Kernighan, B. W. and Weinberger, P. J. (1988): The AWK programming language. Addison-Wesley.

Angel, E. and Bellman, R. (1960): Dynamic Programming and Partial Equations. Academic Press, New York.

Allen, T. F. H. and Starr, T. B. (1982): Hierarchy. University of Chicago Press, Chicago, USA.

Atkins, P. W. (1986): Physical Chemistry. Oxford Univ. Press, 3rd ed.

Baccini, P. and Brunner, P. H. (1991): Metabolism of the Anthroposphere. Springer, Berlin, Germany.

BBA Biologische Bundesanstalt für Land- und Forstwirtschaft: Höchstmengenliste. 3rd ed. 1989, and Rückstandsliste, 4th ed. 1988. Braunschweig, Germany.

Bear, J. (1972): Dynamics of Fluid in Porous Media. Elsevier, Amsterdam, NL.

Bear, J. (1979): Hydraulics of Groundwater. McGraw-Hill, New York.

Benedict, B. A. (1981): Modeling of Toxic Spills into Waterways. In *Hazard Assessment of Chemicals - Current Developments Vol. 1',* Saxena, J. and Fisher, F. (Eds.), Academic press, New York.

Benzler J. H., Finnern H., Müller, W. Roeschmann G., Will K. H. and Wittmann, O. (1982): Bodenkundliche Kartieranleitung. Bundesanstalt für Geowissenschaften und Rohstoffe und Geologische Landesämter in der Bundesrepublik Deutschland (Eds.), Hannover, Germany.

BGBl Bundesgesetzblatt (1986): Verordnung über Trinkwasser und über Wasser für Lebensmittelbetriebe (Trinkwasserverordnung - TrinkwV) vom 22. Mai 1986 (Bundesgesetzblatt BGBl, Teil I, 760–773, 28.5.1986). German Law.

Bidleman, T. F. (1988): Atmospheric Processes. *Environ. Sci. Technol.* 22, 361–367.

Braun, M. (1983): Differential Equations and Their Applications. 3rd ed. Springer, New York, Heidelberg, Berlin.

Briggs, G., Bromilow, R. and Evans, A. (1982): Relationships Between Lipophilicity and Root Uptake and Translocation of Non-ionised Chemicals by Barley. *Pestic.Sci.* 13, 495–504.

Briggs, G., Bromilow, R., Evans, A. and Williams, M. (1983): Relationships Between Lipophilicity and the Distribution of Non-ionised Chemicals in Barley Shoots Following Uptake by the Roots. *Pestic.Sci.* 14, 492–500.

Briggs, G., Rigitano, R., and Bromilow, R. (1987): Physico-chemical Factors affecting Uptake by Roots and Translocation to Shoots of Weak Acids in Barley. *Pestic.Sci.* **19**, 101–112.

Brodsky, J. (1986): Zusammenhang von Molekülstruktur und Retention in "reversed-phase"-HPLC-Systemen bei Chlorbenzolen und Polychlorbiphenylen. Doctoral thesis at the faculty of natural science and mathematics of the University of Ulm, Germany.

Bromilow, R. H. and Chamberlain, K. (1995): Principles governing Uptake and Transport of Chemicals. In *Plant Contamination. Modeling and Simulation of Organic Chemicals Processes.* Trapp, S. and Mc Farlane, J. C. (Eds.), Lewis Pub., Boca Raton, 37–68.

Brüggemann, R., Trapp, S. and Matthies, M. (1991): Behavior Assessment of a Volatile Chemical in the Rhine River. *Environ. Toxicol. Chem.* **10**, 1097–1103.

Brüggemann, R. and Altschuh, J. (1991): Validierung von Abschätzmethoden für physikalisch-chemische Eigenschaften organischer Substanzen. GSF-report 34/91, Neuherberg, Germany.

Brüggemann, R., Münzer, B. and Altschuh, J. (1992): Abschätzung von expositionsrelevanten Substanzeigenschaften. GSF-report 5/92, Neuherberg, Germany.

Burns L. A., Cline D. M. and Lassiter R. R.(1992): Exposure Analysis Modeling System (EXAMS): User Manual and System Documentation. EPA-600 /3-82-023, Environmental Protection Agency, Athens, Georgia, USA.

Butcher, J. C. (1987): The Numerical Analysis of Ordinary Differential Equations: Runge Kutta and General Linear Methods. John Wiley and Sons, New York.

Calamari, D., Tremolada, P., Di Guardo, A. and Vighi, M. (1994): Chlorinated Hydrocarbons in Pine Needles in Europe: Fingerprint for the Past and Recent Use. *Environ. Sci. Technol.* **28**, 429–434.

Chiou, C. T., Peters, J. L. and Freed, V. H. (1979), *Science* **206**, 831.

Churchill and Veith are cited in: Lyman, W., Reehl, W. and Rosenblatt, D. (1982): Handbook of Chemical Property Estimation Methods. McGraw-Hill, New York.

Colborn, T., Dumanoski, D. and Peterson Meyers, J. (1996): Our Stolen Future. Dutton, New York, USA.

Cowan, C., Mackay, D., Feijtel, T., de Meent, D. van, Di Guardo, A., Davies, J. and Mackay, N. (Eds., 1996): The Multi-Media Fate Model: A Vital Tool for Predicting the Fate of Chemicals. Setac Press, Pensacola, Florida, USA.

Crank, J. (1970): The Mathematics of Diffusion. Oxford University Press, London, GB.

Doucette, W. J. and Andren, A. W. (1988): Estimation of Octanol/Water Partition Coefficients: Evaluation of six Methods for Highly Hydrophobic Aromatic Hydrocarbons. *Chemosphere* **17**(2), 345–359.

Drescher-Kaden, U., Brüggemann, R., Matthies, M. and Matthes, B. (1990): Organische Schadstoffe im Klärschlamm. Ecomed, Landsberg am Lech, Germany.

Dyck, S. and Peschke, G. (1983): Grundlagen der Hydrologie. Ernst and Sohn, Berlin, Germany.

ECETOC (1994): Environmental Exposure Assessment. Technical Report No. 61. ECETOC Brussels, Belgium.

Enquete(1993): Enquete-Kommission "Schutz des Menschen und seiner Umwelt – Bewertungskriterien und Perspektiven für umweltverträgliche Stoffkreisläufe in der Industriegesellschaft". Prereport 30.9.1993, Bundestagsdrucksache 12/5812.

EU (1996): Technical Guidance Documents in Support of The Commission Directive 93/67/EEC on Risk Assessment For New Notified Substances and the Commission Regulation (EC) 1488/94 on Risk Assessment For Existing Substances. ECB, Ispra, Italia, advanced pre-print-Version.

Feher, J., M. Th. van Genuchten and T. Nemeth (1991): Nitrogen Leaching from Agricultural Soils – A Comparison of Measured and Computer-Simulated Results. In *Water Pollution: Modelling, Measuring and Prediction,* Wrobel, L. C. and Brebbia, C. A. (Eds.),

Computational Mechanics Publications, Southhampton, GB, and Elsevier Applied Science, Amsterdam, NL.

Fischer, H. B. et al. (1979): Mixing in Inland and Coastal Waters. Academic Press, New York.

Gates, D. M. (1980): Biophysical Ecology. Springer, New York.

GDCH Gesellschaft Deutscher Chemiker (Eds., 1988): Altstoffbeurteilung. GDCH-Geschäftsstelle, P.O.box 900440, Varrentrappstraße 40-42, Frankfurt am Main, Germany.

Gear, C. W. (1975): Numerical Initial Value Problems in Ordinary Differential Equations. Prentice Hall Inc., Englewood Cliffs, N.J.

Genuchten, M. Th. van and Alves, W. J. (1982): Analytical Solutions of the One-Dimensional Convective-Dispersive Solute Transport Equation. U.S. Department of Agriculture Technical Bulletin No.1661, 1982.

Graedel, T. E., Hawkins, D. F. and Claxton, L. D. (1986): Atmospheric Chemical Compounds. Academic Press, London, GB.

Haase, G. (1991): Naturraumerkundung und Landnutzung. Beiträge zur Geographie Bd. 34, Akademie Verl. Berlin, Germany.

Hamaker J. W. and J. M. Thompson (1972): Adsorption. In *Organic Chemicals in the Soil Environment*, Goring, C. and Hamaker, J. (Eds.), marcel dekker, New York.

Hermann, R. (1984): Einführung in die Hydrologie. Teubner, Stuttgart, Germany.

Hollemann, A. F. (1976): Lehrbuch der anorganischen Chemie / Hollemann-Wiberg. 81.-90. ed. (Ed. E. Wiberg), de Gruyter, Berlin.

Hsu, F. C., Marxmiller, R. L., Yang, A. S (1991): Study of root uptake and xylem translocation of cinmethlyn and related compounds in detopped soybean roots using a pressure chamber technique. *Plant Physiology* 93, 1573-1578.

Huber, B. (1956): Die Saftströme der Pflanze. Springer, Berlin, Germany.

IKSR – Internationale Kommission zum Schutze des Rheins gegen Verunreinigung (1987): Tätigkeitsbericht 1986. Koblenz, Germany.

Isnard, P. and Lambert, S. (1988): Estimating Bioconcentration Factors from Octanol-Water Partition Coefficients and Aqueous Solubility. *Chemosphere* 17(1), 21–34.

Jacob, F., Jäger, E. and Ohmann, E. (1987): Botanik. Gustav Fischer, Stuttgart, Germany, 3rd ed.

Jacquez, J. A. (1972): Compartmental Analysis in Biology and Medicine. Elsevier, Amsterdam, NL.

Jost, W. (1960): Diffusion in solids, liquids and gases. Academic Press, 1960, 3rd ed.

Junge, C. E. (1975): Basic Considerations About Trace Constituents in the Atmosphere as related to the Fate of Global Pollutants. In Suffett, I.H. (Ed.): Fate of Pollutants in the Air and Water Environments. Wiley, New York, 7–26.

Jury, W. A., Spencer, W. F., Farmer, W. J. (1983): Behavior Assessment Model for Trace Organics in Soil: I. Model Description. *J. Environ. Qual.* 12(4), 558–564; + Erratum (1987) 16(4), 448.
See also:
Jury W. A., Spencer, W. F. and Farmer, W. J. (1983): Use of Models for Assessing Relative Volatility, Mobility, and Persistence of Pesticides and Other Trace Organics in Soil Systems. In *Hazard Assessment of Chemicals*, Saxena, J. (Ed.) Vol. 2, Academic Press, New York.

Karickhoff, S. W. (1981): Semi-Empirical Estimation of Sorption of Hydrophobic Pollutants on Natural Sediments and Soils. Chemosphere 10, 833–846.

Kenaga , E. E. and Goring, C. A. I. (1980), cited in Lyman et al. (1990).

Kerler, F. and Schönherr, J. (1988): Permeation of lipophilic chemicals across plant cuticles: prediction from partition coefficients and molar volumes. *Arch. Environ. Contam. Toxicol.* 17, 7–12.

KHR (1991): Rheinalarmmodell Version 2.0 – Kalibrierung and Verifikation. Report # II-4 of the KHR. ISBN 90-70980-12-6.

Kinzelbach, W. (1992): Numerische Methoden zur Modellierung des Transports von Schadstoffen im Grundwasser. Oldenbourg, Munich, Germany, 2nd ed.

Kjeller, L.-O., Jones, K. C., Johnston, A. E. and Rappe, C. (1991) *Environ. Sci. Technol.* **25**, 1619.

Klöpffer, W., Rippen, G. and Frische, R. (1982): Physico-Chemical Properties as Useful Tools for Predicting the Environmental Fate of Organic Chemicals. *Ecotox. Environ. Safety* **6**, 294–301.

Klöpffer, W. (1994): Environmental Hazard - Assessment of Chemicals and Products. Part II: Persistence and Degradability of Organic Chemicals. *ESPR - Environ. Sci. & Pollut. Res.* **1**(2), 108–116.

Komoßa, D., Langebartels, C. and Sandermann, H. (1995): Metabolic Processes for Organic Chemicals in Plants. In *Plant Contamination. Modeling and Simulation of Organic Chemicals Processes.* Trapp, S. and Mc Farlane, J. C. (Eds.), Lewis Pub., Boca Raton, 69–103.

Korte, F. (1992): Lehrbuch der ökologischen Chemie: Grundlagen und Konzepte für die ökologische Beurteilung von Chemikalien. Thieme, Stuttgart, Germany, 3rd ed.

LAI Länderausschuß für Immissionsschutz (1992): Krebsrisiko durch Luftverunreinigungen. Ministerium für Umwelt, Raumordnung und Landwirtschaft des Landes Nordrhein-Westfalen (Ed.), Germany.

Lapidus, L. and Schiesser, W. E. (1975): Numerical Methods for Differential Systems. Academic Press, New York.

Lapidus, L. and Pinder, G. F. (1982): Numerical Solution of Partial Differential Equations in Science and Engineering. John Wiley and Sons, New York.

Leeuwen, C. J. van and Hermens, J. L. M. (Eds., 1995): Risk Assessment of Chemicals: An Introduction. Kluwer Acad. Press, Dordrecht, NL.

Lepper, P., Klein, M., Knoche, H. and Herrchen, M. (1994): Regionale ökotoxikologische Risikoabschätzung - Möglichkeiten und Grenzen. In *Ecoinforma Vol. 7*, Totsche, K. and Matthies, M. (Eds.), 383–390.

Liss, P. S. and Slater, P. G. (1974): Flux of Gases across the Air-Sea Interface. *Nature* **274**, 181.

LWA Landesanstalt für Wasser und Abwasser Nordrhein-Westfalen (1987): Gewässergütebericht 1986. Adress: LWA Auf dem Draap 25, Düsseldorf, Germany.

Luenberger, D. G. (1979): Introduction to Dynamic Sytems. John Wiley and Sons, New York.

Lyman, W., Reehl, W. and Rosenblatt, D. (1990, eds.): Handbook of Chemical Property Estimation Methods. McGraw-Hill, New York.

Mackay, D. (1979): Finding fugacity feasible. *Environ. Sci. Technol.* **13**, 1218–1223.

Mackay, D. and Paterson, S. (1981): Calculating Fugacity. *Environ. Sci. Technol.* **15**(9), 1006–1014.

Mackay, D. and Yeun, A. T. K. (1983): Mass Transfer Coefficients for Organic Solutes from Water. *Environ. Sci. Technol.* **17**, 211.

Mackay, D. and Yeun, A. T. K. (1980): Volatilization Rates of Organic Contaminants from Rivers. Walter Poll. Res. J. of Canada 15, 83.

Mackay, D., Paterson, S., Cheung, B., and Nealy, W. (1985): Evaluation of the Environmental Behavior of Chemicals with a Level III Fugacity Model. *Chemosphere* **14**(3/4), 335–375.

Mackay, D., Paterson, S. and Schroeder, W. H. (1986): Model Describing the Rates of Transfer Processes of Organic Chemicals between Atmosphere and Water. *Environ. Sci. Technol.* **20**, 810–816.

Mackay, D. (1991): The Fugacity Approach. Multimedia Environmental Models. Lewis Pub., Michigan, USA.

Mackay, D., Paterson, S. and Shiu, W. Y. (1992): Generic Models for Evaluating the Regional Fate of Chemicals. *Chemosphere* **24**(6), 695–717.

Matthies, M. (1991): Expositionsmodelle. UWSF-Z. *Umweltchem. Ökotox.* **3**(1), 37–41.

Matthies, M., Behrendt, H. and Münzer B. (1987): EXSOL Modell für den Transport und Verbleib von Stoffen im Boden. GSF-Report 23/87 Neuherberg, Germany.

Matthies, M., Brüggemann, R., Trapp, S., Münzer, B., Staehle, B. and Trapp, M. (1992): Methoden zur Früherkennung von Boden- und Gewässerbelastungen. Teil II: Expositionsanalyse für Chemikalien in Gewässern. Report to the German Ministry of Research and Technology, Germany.

Matthies, M., Brüggemann, R. and Münzer, B. (1992a): Screening Assessment Model System SAMS. Prepared for OECD Chemicals Division, Paris (France) and Umweltbundesamt, Berlin, Germany.

Matthies, M., Wagner, J.-O. and Koormann, F. (1997): Combination of Regional Exposure Models for European Rivers with GIS Information. In *Information and Communication in Environmental and Health Science.* Ecoinforma '97 (in press).

Mauseth, J. D. (1988): Plant Anatomy. The Benjamin/Cummings Publishing Company, Menlo Park, California.

Mazjik, A. van (1987): Die Dispersion von Stoffen im Rhein und ihre Konsequenzen für die Gewässerschutzpolitik. Bericht von der 11. Arbeitstagung der internationalen Arbeitsgemeinschaft der Wasserwerke im Rheineinzugsgebiet; NL-1005 AD Amsterdam, P.O.box 8169.

McCrady, J. and Maggerd, S. P. (1993) *Environ. Sci. Technol.* **27**, 343.

Mc Farlane, C., Pfleeger, T. and Fletcher, J. (1990) *Environ. Toxicol. Chem.* **9**, 513.

McLachlan, M. S. (1992): Das Verhalten hydrophober chlororganischer Verbindungen in laktierenden Rindern. Doctoral Thesis at the University of Bayreuth, FRG, and personal communication.
See also:
McLachlan, M. (1994): Model of the Fate of Hydrophobic Contaminants in Cows. *Environ. Sci. Technol.* **28**, 2407–2414.

Millington, R. J. and Quirk, J. M. (1961): Permeability of Porous Solids. *Trans. Faraday Soc.* **57**, 1200–1207.

Monin and Yaglom are cited in: Trapp, S. and Brüggemann, R. (1988): Untersuchung der Ausgasung leichtflüssiger Substanzen aus mitteleuropäischen Fließgewässern mit dem Fließgewässermodell EXWAT. Deutsche Gewässerkundliche Mitteilungen **32**(3), 79–85

Nirmalakhandan, N. N. and Speece, R. E. (1988): QSAR Model for Predicting Henry's Constant. *Environ. Sci. Technol.* **22**, 1349–1357.

OECD (1992): Application of Simple Models for Environmental Exposure Assessment. Report of the Workshop 11.-12. Dec. 1992 Berlin, Germany.

Persicani, D. (1993): Atrazine leaching into groundwater: comparison of five simulation models. *Ecological Modelling* **70**, 239–261.

Reichert, P. and Wanner, O. (1987): Simulation of a severe case of pollution of the Rhine river. In *Proceedings of the Twelfth Congress of the International Association of Hydraulic Research;* Littleton: Water Resources; 239–244.

Richter, J. (1990): Models for Processes in Soil. Catena, Reiskirchen, Germany.

Richtmeyer, R. D. and Morton, K. W. (1967): Difference Methods for Initial Value Problems. Wiley-Interscience, 2nd edition, New York.

Riederer, M. (1995): Partitioning and Transport of Organic Chemicals between the Atmospheric Environment and Leaves. In *Plant Contamination. Modeling and Simulation of Organic Chemicals Processes.* Trapp, S. and Mc Farlane, J. C. (Eds.), Lewis Pub., Boca Raton, 153–190.

Rigitano, R., Bromilow, R., Briggs, G. and Chamberlain, K. (1987): Phloem Translocation of Weak Acids in Ricinus Communis. *Pestic.Sci.* **19**, 113–133.

Rippen, G. (1996): Handbuch der Umweltchemikalien. ecomed, Landsberg am Lech, Germany (permanently actualized).

Rordorf, B. F. (1992): personal communication

Rykiel, E. J. (1996): Testing ecological models: the meaning of validation. *Ecological Modelling* **90**, 229–244.

Sachs, L. (1992): Angewandte Statistik. 7th ed., Springer, Berlin.

Samenwerkende Rijn- en Maaswaterleidingbedrijven (1987): De samenstelling van het Rijnwater in 1986 en 1987. RIWA P.O.box 8169; 1005 AD Amsterdam, NL.

Schachtschabel, P., Blume, H.-P., Brümmer, G., Hartge, K.-H. and Schwertmann, U. (1989): Scheffer/Schachtschabel Lehrbuch der Bodenkunde. 12th ed., Enke-Verlag, Stuttgart, Germany.

Scheil, S., Baumgarten, G., Matthies, M., Reiter, B., Schwartz, S., Trapp, S. and Wagner, J.-O. (1995): An Object-Oriented Software for Fate and Exposure Assessments. *ESPR - Environ. Sci. & Pollut. Res.*, Vol. 2 No. 4, 1995, ISSN 0944-1344, 238–241.

Schlegel, H.G. (1981): Allgemeine Mikrobiologie. Thieme, Stuttgart, Germany.

Schroll, R and Scheunert, I. (1992) *Chemosphere* **24**, 97.

Schwarzenbach, R. and Westall, J. (1981): Transport of Nonpolar Organic Compounds from Surface Water to Groundwater: Laboratory Sorption Studies. *Environ. Sci. Technol.* **15**, 1360–1367.

Schwarzenbach, R., Gschwend, P. und Imboden, D. (1993): Environmental Organic Chemistry. John Wiley and Sons, New York.

Shampine, L. F.; Gordon, M. K. (1975): Computer Solutions of Ordinary Differential Equations – The Initial Value Problem. Freeman, San Francisco.

Southworth, G. R. (1979): The Role of Volatilization in Removing Polycyclic Aromatic Hydrocarbons from Aquatic Environments. *Bull.Environ. Contam. Toxicol.* **21**, 507–514.

Strang, G. and Fix, G. J. (1973): An Analysis of the Finite Element Method. Prentice Hall, Englewood Cliffs/NJ.

Suzuki, T. and Kudo, Y. (1990): Automatic log P estimation based on combined additive modeling methods. *Journal of Computer-Aided Molecular Design* **4**, 155–198.

TA-Luft (1986): Technische Anleitung zur Reinhaltung der Luft; In Vogl, Heigl, Schäfer (eds.): Handbuch des Umweltschutzes, Vol. II - 2, appendix 3.1. ecomed Landsberg am Lech, Germany.

Teuber, W. and Wander, K. (1987): Fließzeiten im Rhein aus Flügelmessungen. Bundesanstalt für Gewässerkunde BfG-0392 Koblenz, Germany.

Thompson, N. (1983): Diffusion and Uptake of Chemical Vapour Volatilising from a Sprayed Target Area. *Pestic. Sci.* **14**, 33–39.

Tinsley, I. (1979): Chemical Concepts in Pollutant Behaviour. John Wiley and Sons, New York.

Trapp, S., Matthies, M., Scheunert, I. and Topp, E. M. (1990): Modeling the Bioconcentration of Organic Chemicals in Plants. *Environ. Sci. Technol.* **24**(8), 1246–1251.

Trapp, S. and Pussemier, L. (1991): Model Calculations and Measurements of Uptake and Translocation of Carbamates by Bean Plants. *Chemosphere* **22**(3-4), 327–339.

Trapp, S. (1995): Model for Uptake of Xenobiotics into Plants. In *Plant Contamination. Modeling and Simulation of Organic Chemicals Processes*. Trapp, S. and Mc Farlane, J. C. (Eds.), Lewis Pub., Boca Raton, 107–151.

Trapp, S. and Mc Farlane, J. C. (1995, Eds.): Plant Contamination – Modeling and Simulation of Organic Chemicals Processes. Lewis Pub., Boca Raton, USA. ISBN 0-56670-078-7. with contributions from S. Trapp, C. Mc Farlane, R. H. Bromilow, K. Chamberlain, D. Komoßa, C. Langebartels, H. Sandermann jr., M. Riederer, S. Paterson, D. Mackay, M. Matthies and H. Behrendt.

Trapp, S., Rantio, T. and Paasivirta, J. (1994): Fate of Pulp Mill Effluent Compounds in a Finnish Watercourse. *Environ. Sci. & Pollut. Res.* 1(4), 246–252.

Trapp, S. and Harland, B. (1995): Field test of Volatilization Models. *Environ. Sci. & Pollut. Res.* 2(3), 164–169.

Trapp, S. and Matthies, M. (1995): Generic one Compartment Model for the Uptake of Organic Chemicals into Foliar Vegetation. *Environ. Sci. Technol.* 29, 2333–2338; and *Erratum Environ. Technol.* 30, 360.

Trapp, S. and Matthies, M. (1996): Dynamik von Schadstoffen – Umweltmodellierung mit CemoS. Eine Einführung. Springer, Heidelberg, Germany.

Trapp, S. and Matthies, M. (1997): Modeling Volatilisation of PCDD/F from Soil and Uptake into Vegetation. *Environmental Science and Technology* 31(1), 71–74.

Trenkle, R. and Münzer, B. (1987): EXAIR – Analytisches Transportmodell für die atomosphärische Mischungsschicht. GSF-report 31/87, Neuherberg, Germany.

Turner, M. and Gardner, R. H. (Eds., 1991): Quantitative Methods in Landscape Ecology. Ecological Studies Vol. 82, Springer, New York.

UBA Umweltbundesamt (1995): Environmental Data Germany 1995. Umweltbundesamt Dept. I 4.3 'Environmental Reporting, Environmental Statistics', Bismarckplatz 1, 14 193 Berlin, Germany.

US-EPA United States Environmental Protection Agency (1996): Soil Screening Guidance: Technical Background Document. EPA/540/R95/128.

Veerkamp, W. and Wolff, C. (1996): Developments and Validation Criteria. *ESPR - Environ. Sci. & Pollut. Res.* 3(2), 91–95.

Visser, W. J. F. (1993): Contaminated land policies in some industrialized countries. Technische Commissie Bodembescherming (technical soil protection committee), The Hague, NL.

Vogt, S. (1980): Vierparametrige Ausbreitungsstatistik als Berechnungsgrundlage der langzeitigen Schadstoffbelastung in der Umgebung eines Emittenten, KfK Karlsruhe, Germany.

Vvedenski, D. (1993): Partial Differential Equations with Mathematica. Addison-Wesley Publishing Company, Reading, MA.

Wagner, J.-O. and Matthies, M. (1996): Guidelines for Selection and Application of Fate and Exposure Models. *ESPR - Environ. Sci. & Pollut. Res.* 3(1), 47–51.

Walther, H. and Lieth, H. (1967): Klimadiagramm Weltaltas. Fischer, Jena, Germany.

Weast, R. C. and Astle, Ph. D. (Editors) (1983): CRC Handbook of Chemistry and Physics. CRC Press, Boca Raton, Florida, 63th Edition.

Welsch-Pausch, K., Umlauf, G. and McLachlan, M. S. (1995): Determination of the Principal Pathways of Polychlorinated Dibenzo-p-dioxins and Dibenzofurans to Welsh Ray Grass *(Lolium multiflorum). Environ. Sci. Technol.* 29, 1090–1098.

Westrich, B. (1988): Fluvialer Feststofftransport – Auswirkung auf die Morphologie und Bedeutung für die Gewässergüte. Oldenbourg, Munich, Germany.

Whelpdale, D. M. (1982): Wet and Dry Deposition. In Georgii, H. W. and Jaeschke, W. (eds.): Chemistry of the Unpolluted and Polluted Troposphere. D. Reidel Pub.

Whitman W. G. (1923): A preliminary experimental confirmation of the two-film theory of gas adsorption. *Chem.Metall.Eng.* 29, 146–148.

Wolff, C. J. M. and van der Heijde, H.B. (1982): A Model to Assess the Rate of Evaporation of Chemical Compounds from Surface Waters. *Chemosphere* 11(2), 103–117.

Yalkowski, S. H. and Valvani, S. C. (1980), *J.Pharm.Sci.* 69, 912–922.

CemoS User's Manual

For Program Version 1.10

Guido Baumgarten
Bernhard Reiter
Sven Scheil
Stefan Schwartz
Jan-Oliver Wagner

Table of Contents

Chapter 1:
Tutorial

This tutorial was written to introduce new and inexperienced users to the operation of CemoS. It describes step by step a complete run of the program system. It is not the intention of the tutorial to explain every function provided by CemoS. This chapter should not take the place of the other chapters, but should enable new users to familiarize themselves with the program quickly.

After launching CemoS from the command line, the intro screen appears. It contains an information dialog, showing the CemoS version number and the authors' addresses.

Before selecting a model, it is important to collect and evaluate all given (available) data about the environmental segment you would like to simulate. CemoS cannot do this for you. After you have done this, you can select the model that best fits your problem. To do this you have to select the model from the →Submenu Active Model [Page 199], which is nested in the model menu. The Water model should be used for the tutorial. The name of the current model is shown in the status window on the desktop. Before moving on to the next paragraph, ensure that the name of the current model is Water.

The next step in a CemoS session is to select a chemical. This you do by selecting the →Menu command Load Substance [Page 198]. The dialog that appears shows the content of the current substance database. CemoS refers to a database called STDSUB.DAB which is stored in the current directory. You can manipulate the substance database by using the commands of the →Submenu Database [Page 198]. The next step is to select the chemical pentachlorophenol. One way to do this is to double click the name of the chemical. Another possibility to is to search in the database for certain letters in a substance name with the button Search. The chemical data will now be retrieved from the database and the name of the current chemical is shown in the status window. Furthermore, a hint appears which informs us that two model parameters may have to be entered. Such tips always appear when the chemical is changed and if model parameters are indirectly dependent on the substance data, but not automatically calculated. At this point we can pass over this hint.

Before the simulation can be started, the model parameters have to be entered. In order to open the input dialog you have to select → Menu command Parameter input [Page 199]. The input fields for the K_d and volatility rate contain the value 'unknown'. In order to start the simulation, you have to enter sensible values for these two parameters. One way to do this is to manually enter values that you have evaluated during your preconsideration phase. For the tutorial we choose another way: You can use → Estimation functions [Page 260]. To do this select the input field of K_d (e.g. by simultaneously pressing <Alt> and <d>) and then activate the Estimate button. As you can see, the Estimation button is only available when the K_d input field is selected. The Estimation button opens a dialog which contains a list of available Estimation functions. In our case is only one Estimation function available. For detailed information about the Estimation function use the Details button at the bottom of the list. Now activate the estimation function by double clicking its name. After the dialog is closed in the K_d input field, the estimated value appears. This value is linked to the value of the pH by using the Estimation function. This means that changing the pH-value will automatically update the K_d value. You can test this automatic update by changing the default value (6) of the pH to a different value (7). This link is active until the user enters a new value in the K_d input field manually.

In order to complete the set of parameters, activate the first Estimation function for the parameter volatility rate. You are now ready to start the simulation and can press key <F9> or use the → Menu item Start simulation [Page 200]. The result windows will soon appear on the desktop.

You can change the order of the result windows by clicking the right mouse button or by pressing <F6>. If you are using an EGA or VGA graphic adapter, you can increase the number of rows that are displayed by using → Menu item Lines [Page 201]. Now you can place several windows on the desktop without overlap. A list of the available high-resolution graphic plots can be activated with → Menu item Show graphics [Page 200]. You can activate a hires plot by double clicking its name. The hires mode can be closed by pressing any key.

Let us assume that you need a print-out of the simulation, as well as a portrayal of the water diagram with gnuplot. From the menu Program you select here the → Menu item Export As [Page 196]. You then enter an appropriate file name, e.g. TEST. All data are now saved under TEST.MIF in MIF format.

Now leave CemoS without saving the scenario. Enter the command REPORT TEST.MIF TEST.PRT at the DOS prompt. This command generates the file TEST.PRT which contains all information of the model run. You can now send this file to your printer or use it in your word processor.

Another feature of the export interface is that you can use the simulation results with other applications. For the following example it is assumed that you have installed gnuplot on your computer. The command PLOT TEST.MIF extracts the required data from the TEST.MIF file and send them to gnuplot.

An advantage of using gnuplot for your plots is its ability to convert the plot to postscript or to LaTeX commands.

You have now reached the end of the tutorial and we hope you have gained an impression of working with CemoS. Most of the remaining questions should be answered in the online help or in the manual. The index of the online help can be invoked by pressing <Shift> + <F1>.

We wish you a lot of fun and success in working with CemoS!

Chapter 2:
Desktop

2.1
Help System

CemoS provides the user with a context-sensitive help system. At any stage of
the program you can receive help on the relevant topic by pressing <F1>. The
only exception to this are high- resolution graphics. Special help topics can be
accessed via →Menu Help [Page 204].

A help topic gives hints in text form about a certain subject. In all help texts
some words are highlighted. These are cross-references; by activating them
with <Tab> and <Enter> you will be led to the relevant subject. A new help
text is shown and the keyword you clicked on is explained. Every help text has
between one and three standard cross-references at the top. With these you can
move through the tree structure of the help system. The reference "Up" leads
you to the upper level, which contains the current term as a subtheme. The
references "Right" and "Left", if available, lead you to related topics belonging
to the same main topic.

As you can see from the status line at the bottom of the screen, there are
further keys to enable you to move within the help system. <F1> shows you
this help text so that you can get help on the system whenever you need it.
<Ctrl> + <F1> shows you the context table of the help system. From there you
can move to specific subjects. The index can be shown by pressing <Shift> +
<F1>. If you are searching for a certain keyword, you can easily find it in the
alphabetically ordered index and can get the available help text on it. <Alt>
+ <F1> brings you back to the latterly displayed help text. The sequence of
the topics shown is recorded from the moment you start the help system, so
that you can always return to the first one. As soon as you exit from the help
system, the records are deleted and will be restarted by calling the help system
again.

You can exit from the help system via the <Esc> key or by clicking the close
field at the top left of the window.

2.2
Using the Mouse

All functions of this program can be controlled using the mouse. A click of the left mouse button opens menus, starts menu commands, presses buttons, selects dialog elements, marks list items and activates windows. Holding the left mouse button down you can move or change the size of windows. By doubleclicking the left button, list items can be marked and at the same time a specific action be started. Switching to the next window can be done by clicking the right mouse button, which is the same action as →Menu item Next Window [Page 203].

2.3
Using the Keyboard

All functions of the program can also be controlled using the keyboard.

The menu can be activated by <F10>. The <Return> key opens a submenu or executes a menu command. You can scroll through the menu with the cursor keys.

Many menu commands have a key name on the right of the menu. These hotkeys execute the corresponding menu command without naving to open the menu.

You can change the active window with the menu command window list or by pressing the number of the window while holding the <Alt> key. The window number can be found at the top right of the window.

In dialogs you can change the active dialog element using the <Tab> key. In input lines or fields you can enter any text or number via the keyboard. With selection fields the status is changed using the cursor keys. With multi-selection fields items can be selected using the space-bar. In list boxes you can move using the cursor keys, <Home>, <End>, <PgUp> and <PgDn>. With multi-selection list boxes you can mark an item with the space-bar. Once a button is selected and therefore highlighted, you can carry out the corresponding action by pressing the return key.

Most dialog elements are given a term with a highlighted character which is henceforth called a contraction. These dialog elements can be selected by pressing <Alt> and the relevant character key. A dialog can be cancelled by pressing <Esc>.

2.4
Menu

The menu is the most important control item in this program. It can be found at the top of the screen. All major commands can be started from the menu, which can be activated by <F10> or a single click.

A detailed description of the menu commands can be found under → Menu Items [Page 196].

2.5
Windows

Windows are rectangular areas on the screen which contain certain information. The active window has a double frame, whereas all others have a single frame. Window operations will henceforth be described using the mouse; you can find a description of how to carry out the same operations with the keyboard under → Menu Windows [Page 203].

The contents of the window are briefly described by a title at the top of the frame. By clicking on this title line and holding the mouse button you can move the window around the screen. The specific window number is at the top right. If you wish to move to a certain window, you can either do this by using the window list or simply pressing <Alt> and the corresponding number of the window. If a window has scrollbars, the window is not able to display the whole content. By using the scrollbars you can be shown the remaining area. By clicking on the arrows you can scroll the window content in single steps. To scroll faster you have to click on the line between the arrows or at the position pointer and move this whilst holding the mouse button.

Some windows can be made larger or smaller. These windows have a small area at the bottom right, which looks like a single frame although the window is active. By clicking on this symbol and holding the mouse button, you can change the size of the window by moving the mouse.

To increase a window to its maximum size quickly, you just have to click the symbol at the top right, if available. If this symbol is clicked a second time, the window returns to its previous size. A double click on the top line of the window frame results in the same function.

Two windows are displayed permanently, the → Status window [Page 188] and the → Protocol window [Page 189].

Similar to the input dialogs, the model result windows will be explained separately within the description of each model. Most of the model results tables are presented in a → Table window [Page 189] and in one or more → Graphic windows [Page 189].

2.5.1
Status Window

This window is permanently shown on the screen. It displays the following information:

- the substance database currently open
- the number of substances in the current substance database
- the currently entered or loaded substance
- the currently selected model
- the loaded scenario for the selected model.

2.5.2
Protocol Window

The protocol window collects all hints and warnings which may appear whilst working on a simulation. These could be warnings caused by ignoring suggested parameter ranges or hints and warnings due to applied estimation functions. Additionally, the window contains further information concerning unknown parameters or messages resulting from the execution of a model.

If the results of a simulation differ from the expected values, possible error sources should first be looked for within this window. Once the source of a warning/hint has been removed, the information will disappear from the protocol window.

2.5.3
Table Window

Most models use a table window to display the result output. Here, for example, a concentration could be portrayed in relation to the distance. A table window can display a table with an unlimited number of rows and columns. To view all values of the table, the window has two scrollbars. If you move down through the output using the vertical scrollbar, the column titles will keep their position. You thus retain an overview of the values currently being portrayed.

The horizontal scollbar moves the output columnwise, if all columns cannot be displayed at the same time.

In most cases columns from the second one onwards are shown in relation to the first one, which is separated from the others by a double line and cannot be moved horizontally.

2.5.4
Graphic Window

A graphic window displays two columns in a 2-D graph. These have a horizontal scrollbar to move through the whole output; the height is adjusted automatically.

Since graphic windows only work in text mode, the values cannot be displayed exactly. In order to get the exact values you have to refer to the corresponding table window. If you wish to view a more exact graph quickly, you can display the same graph in a high-resolutuion mode. This you start with →Menu item Show Graphics [Page 200]. High-resolution graphics require a specific graphic adapter to be installed.

2.6
General Dialog Elements

Dialogs, which can be called up in the same method as →Windows [Page 188], are special windows with additional fields, called dialog elements, where the user can enter input data. All dialog elements appearing in CemoS will be explained in this section.

In addition to these general elements, the dialogs for substance and model parameters possess some common features:

1. All input dialogs have a →Comment field [Page 192] in which annotations to the entered data can be added.
2. All dialogs have a comment button (→Buttons [Page 190]). If you click this button, another dialog opens in which you can enter an annotation to the corresponding value. Furthermore the dialog also contains information about the estimation function used if the value has been estimated.
3. Most input dialogs also have an estimation button (→Buttons [Page 190]). If you select an input line where the value can be estimated, this button becomes selectable (buttons that cannot be selected are shown grey). If you click this button the →Dialog Estimation functions [Page 211] will open. There you can select an estimation function to estimate the value.

2.6.1
Buttons

All dialogs contain buttons whose function it is to carry out a certain action if they are clicked on by the mouse or selected via <Tab> and entered using <Return>.

Nearly all dialogs have at least two standard buttons: the Ok button and the Cancel button. As a substitute for the latter, you can also use the mouse to click on the close symbol at the left top of the window or just press <Esc>.

In addition to these two buttons, most dialogs possess several other buttons for various commands:

Addresses: This button is used in the info dialog to display the address dialog which contains information on how to contact the authors.

All: This button appears in dialogs with list boxes to mark all items in the list with only one click, so that the desired selection can be achieved quickly.

Cancel: This button is used to cancel a dialog and to discard the entered or modified values. After such a cancelation, the program acts as if the dialogs had never been opened. In message windows this button cancels the action that produced the message.

ChDir: This button is used in the →Dialog Change Directory [Page 193] to change to the selected directory. A double-click on the directory name has the same effect.

Comment: This button is used to comment on a value in the input dialogs for substance data and model parameter. This button can only be selected if the active value can be commented upon.

Continue: When loading and deleting substances, this button is used to search for the next matching substance following a search process.

Delete: This button is used in the →Dialog Delete Substance [Page 216] to close the dialog and to delete the marked substance. Similar to the load button, this button has the same function as the Ok button, but has a more fitting name.

Details: This button is used in the → Dialog Estimation functions [Page 211] to show you a help text with detailed information about the selected estimation function.

Estimate: This button is used to estimate a value in the input dialogs of substance data and model parameters. The button can only be selected if the active value can be estimated.

Fill: This button is used for the input of parameters in the → Buckets [Page 227] model. It facilitates the input of tables. By pressing this button, the value of the current position will be transferred to all of the following rows of the same column.

Find: This button is used in dialogs that expect the input of a filename. By pressing this button a → File dialog [Page 193] is opened in which you can select an existing filename rather than having to enter the whole name.

Load: This button is used in → Dialog Load Substance [Page 215] to close the dialog and to load the marked substance. It has the same function as the Ok button, but has a more fitting name.

No: This button is used in message windows to reject a decision.

Ok: This button is used to close a dialog and to keep the entered or modified data. Further actions can be executed dependent on the dialog.

Open: This button is used in a → File dialog [Page 193] to open the specified file.

Repeat: This button is used when a system or DOS error occurs in order to repeat the failed action.

Reset: This button is used in the → Dialog Change Colors [Page 194] to change back the colors to the default colors of CemoS.

Revert: This button is used in the → Dialog Change Directory [Page 193] to return to the directory which was present when the dialog was opened.

Save: This button is used to save files and scenarios. It ends the dialog and saves the relevant data.

Search: When loading or deleting substances, this button is used to open → Dialog Search Substance [Page 216].

Yes: This button is used in message windows to accept a decision.

2.6.2
Input Lines

Input lines are dialog elements in which a one-line text or a value can be entered. They are used in CemoS to specify, for example, the substance data and model parameters.

2.6.3
Input Fields

Input fields are multi-line dialog elements in which multi-line, longer texts can be entered. These fields behave in a similar way to simple editors. The only key that cannot be used for inputting is <Tab>, since this key is used to move the active dialog element to the next one. In CemoS input fields are used, amoungst other things, to enter a comment on the data with the input of substance data and model parameters (see also → Comment field [Page 192]).

2.6.4
Comment Field

Comment fields are special input fields and are available in all substance data and model parameter input dialogs. They enable you to enter comments on the entered data sets. Comments enable you to explain the entered values, e.g. which literature they are taken from and who inserted the values. With the substance data a comment has to be entered so that the substance can be saved. The importance of this becomes obvious when you attempt to find the cause of faulty values which give you incorrect results (see also → Input fields [Page 191]).

2.6.5
Selection Fields

Selection fields are dialog elements where you can choose between various different items. Selection fields can be recognized by the brackets that precede them. The position between them may also be marked. There are two types of selection fields:

The single selection fields only allows one item to be selected. If you select another item, the previously marked one will be deselected. Such fields can be recognized by their round brackets and marking point. An example is the protolysis field of the substance data.

Multi-selection fields allow the choice of any combination of one or more items. It is also possible not to select any of the items. Such fields can be recognized by their square brackets and a marking cross. An example is the selection of compartments in the Level models.

2.6.6
List Boxes

List boxes are dialog elements twhere you can choose an item from a list. If more items are available in the list than the box can display, you can use the scroll bar on the right to move to the desired items.

You also can select an item by double-clicking on it and activating the Ok or an analogous button. An example of a list box is the window list.

2.7
Standard Dialogs

In addition to the dialogs for the input of substance data and model parameters, there are several further general dialogs which will be explained in the following. Special dialogs are explained at places suitable to the topic.

2.7.1
File Dialog

A file dialog is used to specify a new filename or to select one from a list of already existing names.

At the left top the dialog has an input line where you can enter the desired filename. To the right of the input line is a small button with an arrow pointing downwards. If you click this button you open a list of all filenames entered or selected by you since starting the program. Thus you can quickly select a name you have already used.

Below the input line the dialog contains a two-column list with all filenames and subdirectories of the current directory. A directory can be recognized by the backslash which follows it. You can use the DOS-joker in the input line to preselect the filenames in the list. If you select a filename from the list, its name then appears in the input line and you now can carry out the desired actions on the selected file with the buttons. If you carry out an action on a directory, the file list is changed to the selected directory and you can go on with your selection.

To the right of the input line and file list there are two → Buttons [Page 190]. The upper button is either called Ok or Open, depending on the type of dialog. The lower button is the Cancel button.

2.7.2
Change Directory

This dialog is used to change the current working directory and scenario directory. If you open a file dialog, the contents of the current directory will be displayed. If you wish to change this, you have to specify a new current directory in this dialog which is similar to a → File dialog [Page 193]. At the top left of the dialog there is an input line in which you can enter the filename. To the right of this there is a button with an arrow pointing downwards. If you click this button, you open a list of all filenames entered or selected by you since starting the program. Thus you can quickly select a name you already used.

Below the input line there is a directory tree. This tree contains all parent and subdirectories of the current one. Each directory level is indented by two space characters. The current directory is highlighted.

To the right of the directory tree are three → Buttons [Page 190]: Ok, ChDir and Revert.

To change a directory you have to browse through the tree to find the desired one. You can change to a parent or subdirectory by double-clicking it or by highlighting it and clicking the ChDir button.

If you have selected the desired directory, you can activate it by clicking the Ok button. The dialog will then be closed and CemoS will change the current directory to the new one. If you wish to return to the active current directory while selecting a new one, you can click the Revert button. If you cancel the dialog with <ESC>, the current directory will not be changed.

2.7.3
Changing Colors

This dialog is used to change the colors of the program desktop to your individual wishes. At the top left there is a list of the color groups (menus, windows, input dialogs, etc.). Highlight the group in which you wish to make a change.

After having selected a group, a further list containing all items of the chosen group appears to the right of it. If you wish to change the color of a certain item, you have to select it and choose a new foreground and background color from the color table. Below the color table there is a sample text showing the currently selected combination.

Proceed in the same manner for all screen element colors which you wish to change. Confirm your selection by pressing the Ok button.

If you do not wish to apply the new colors after all, you can return to the former ones by leaving the dialog via the Cancel button.

If you wish to return to the original colors of CemoS after having already pressed Ok, you retrieve them by clicking the Reset button.

2.7.4
Window List

This dialog is used to change to a certain window if more than one window is presently shown. It contains a list box with all open windows and two buttons, Ok and Cancel.

You can change to a new window by double-clicking it or by highlighting it and pressing Ok.

2.7.5
Message Window

Message windows are the most simple dialogs. They consist of a text and between one and three buttons. Message windows are used for error messages, hints or requests. Depending on the function of the message window, you click on the buttons to close the window, to repeat an action or to reject/accept a decision.

2.7.6
Information Window

In this dialog information is displayed about the program, the version number and the authors. Additionally you can also go from this dialog to the → Address window [Page 195] and the → System information window [Page 195]. The information window appears at the beginning of the program.

2.7.7
Address Window

This dialog shows the contact addresses of the CemoS developers.

2.7.8
System Information Window

This dialog shows information about value ranges which are the maximum and minimum values that can be entered in the input lines of parameter dialogs.

Chapter 3:
Menu Commands

The menu is the main control element of CemoS. It can be seen at the top of the screen and contains all commands to direct the program execution.

The menu is divided thematically into six parts which are described in the following sections. Only those commands that make sense during the execution of the program are active (selectable).

3.1
Menu Program

The menu commands of the menu program are used to control general functions of CemoS.

3.1.1
Import

This command enables you to import substance and/or model data from an MIF file. See also → Import Data [Page 210].

3.1.2
Export As

This command enables you to export substance and model data as well as the results of a simulation into an MIF file. See also → Export Data [Page 212].

3.1.3
DOS Shell

This command is used to leave the program to a DOS prompt without losing data and settings of CemoS. At the DOS prompt you can run a DOS command or any other program that can handle the remaining memory. To return to CemoS enter EXIT at the DOS prompt.

3.1.4
Change Dir

This command opens the →Dialog Change Directory [Page 193] where you can specify a new working directory.

If you open a →File dialog [Page 193] in CemoS, the current working directory is selected by default.

3.1.5
Exit

This command leaves CemoS and returns to the DOS prompt. If model parameters or substance data have been changed since the last loading or saving, the program asks whether you wish to discard or save the data before leaving the program.

3.2
Menu Substance

This menu contains commands to manage substances. You can enter new substances, load substances from or save substances in databases, create new databases or append a database to the current one.

3.2.1
New Substance

This command opens the →Dialog Substance Data Input [Page 214]. In that dialog you can enter the data of a new substance. This new substance is not saved in the database automatically. If you want to append the new substance to the current database, you have to use the →Menu item Save Substance [Page 198]. CemoS will warn you if there could be a loss of data, i.e. modified, but not saved substances.

3.2.2
Edit Substance

This command opens the →Dialog Substance Data Input [Page 214]. In that dialog you can modify the physico-chemical data of the current substance. Changes are not saved automatically. To save the modifications you have to use the →Menu item Save Substance [Page 198].

3.2.3
Load Substance

This command opens the →Dialog Load Substance [Page 215]. There you can choose a substance from the current database. CemoS will give you a warning if the current substance has not yet been saved. Details can be found under →Load Substance [Page 215].

3.2.4
Save Substance

This command appends the current substance to the current database. This is only possible if the substance has at least a name and a comment. If one of these is missing, a message window will appear. Details can be found under →Save Substance [Page 216].

3.2.5
Delete Substance

This command opens the →Dialog Delete Substance [Page 216]. There you can specify a substance from the current database and delete it. Details can be found under →Dialog Delete Substance [Page 216].

3.2.6
Submenu Database

The commands of this submenu allow you to manipulate substance databases. You can create new or open existing ones, append a database to the current one or delete it. A file dialog will be opened for all commands.

If CemoS finds the file STDSUB.DAB in the program directory, this database will automatically be the current one.

New Database: This command is used to create a new substance database. Details can be found under →New Database [Page 217].

Open Database: With this command you can open an already existing database and make it the current one. Details can be found under →Open Database [Page 217].

Merge Databases: This command is used to merge the current database with another database to create a new one that contains all substances of the two selected databases. Details can be found under →Merge Databases [Page 217].

Delete Database: You can delete a whole substance database using this command. You should only delete those databases that you really do not need any more→ Details can be found under →Delete Database [Page 218].

3.3
Menu Model

This menu contains commands to activate the desired model, input the model parameter, manage scenarios and run the simulation.

3.3.1
Submenu Active Model

Within this submenu you can choose the model you wish to work with. The current model is marked with a tick. If you press <F1> while a certain model is highlighted, you are given technical data on this model. Details can be found under → Select Model [Page 218].

3.3.2
Parameter Input

With this command you reach the input dialogs for the model parameter of the current model. Details can be found under → Model Parameter Input [Page 218].

3.3.3
Submenu Scenario

This submenu offers commands to manage substance data and model parameter sets as scenarios. Further information can be found under → Use of Scenarios [Page 219].

Load scenario: This command is used to load an already existing scenario from an earlier session. Details can be found under → Load Scenario [Page 220].

Save scenario: This command will save the current set of model parameters. This prevents you from having to enter the values again at the next session. Details can be found under → Save Scenario [Page 220].

Save scenario as: Here you can save the scenario under a new name, so that the former one can be kept. Details can be found under → Save Scenario as [Page 220].

Delete Scenario: This command is used to delete an existing scenario created in an earlier session. Details can be found under → Delete Scenario [Page 220].

Scenario directory: This command enables you to set a new standard directory where CemoS will first search for scenarios and where you may save your scenarios. Further information can be found under → Scenario Directory [Page 220].

3.3.4
Submenu Model Configurations

Within this submenu you can load or save the model configurations. These configurations cover the filename of the current substance database and the position and size of windows.

The configurations will be saved in the current working directory. If you select a new model, CemoS will search for a certain configuration file in the actual working directory and, if it exists, will load it.

Load model configurations: This command loads previously saved model configurations.

Save model configurations: This command saves the current model configurations.

These configurations will automatically be loaded if you change the model.

3.3.5
Start Simulation

This command is used to start the simulation of the active model using the current parameter sets. If important parameters are missing, the computation will not be executed and instead an error message will be prompted. All hints occuring during the simulation can be found in the $\rightarrow$ Protocol window [Page 189]. CemoS creates a backup file named "CemoS.SAV" before it starts the simulation. This file contains all important settings and data. If the program crashes while the simulation is running, you can load this backup file at a restart and CemoS has all data and settings it had before the simulation. To ensure that CemoS runs over a LAN, it doesn't save the backup file in the program directory, but first looks in the DOS-environment variable "TMP" and saves the file in that temporary directory. If "TMP" or the directory does not exist, CemoS looks in the environment variable "TEMP". If this variable doesn't exist, either, CemoS takes the directory that was active at the start.

3.3.6
Show Graphics

If the current model offers high resolution graphics, you can display these after a successful simulation with this command. First the $\rightarrow$ Dialog Choose Graphic [Page 212] will be opened where you can choose from one or more graphics. After selecting a graphic, CemoS changes to the high resolution mode and shows the desired graphic. Pressing any button of the keyboard or mouse will bring you back to the program desktop. This command is only executable if the final simulation was successful, the model offers graphics and the graphic adapter of your PC is capable of the high resolution mode.

3.4
Menu Desktop

The menu contains commands which influence the appearance and behavior of the desktop. You may change colors or the number of rows on the screen.

If the program directory contains the configuration file CEMOS.DSK, it will be loaded automatically when starting CemoS. To load your own settings automatically you have to rename another configuration file to CEMOS.DSK.

3.4.1
Colors

This command will open the →Dialog Change Colors [Page 194] where the colors of all desktop elements (menu, window, etc.) can be changed.

3.4.2
Lines

This command changes the number of displayed lines. You can choose between 80×25 (columns $\times$ rows, standard) and 80×43 (EGA) or 80×50 (VGA).

3.4.3
Hints

This command switches on/off the prompting of certain hint-message windows. Hints are brief pieces of information that could be of importance to the user, but which do not alter the program execution (simulation). Switched on, all hints will be shown in message windows, as well as in the protocol window. They will appear in the protocol window even if hints are switched off.

3.4.4
Warnings

This command switches on or off the appearance of certain warning-message windows. With a warning the user can decide whether to ignore it or to find out the cause of this message and stop the current action. If this switch is switched on, a positive answer (that is ignoring) is assumed and a corresponding hint is added in the protocol window.

3.4.5
Comment Watch

This command switches on or off the automatic appearance of a comment dialog after changing a value. If this switch is on, the comment dialog for a value opens if you shift to the next value or want to leave the dialog. This only happens, if the value has been changed and a comment already exists. If the switch is off, you have to remind youself to keep all comments up to date.

3.4.6
Long Numbers

This command switches on or off the long representation of values in input dialogs. If the switch is on, the values in the input fields will be shown as exactly as possible. This also means that you often have to scroll to the right end to see the exponent. If the switch is off, the values will be shortened to fit into the input fields. This means you cannot see the precise value, but therefore you can see the exponent immendiately.

3.4.7
Desktop Load

This command opens a → File dialog [Page 193] to select a configuration file which you wish to load. If at the program start there is a configuration file CEMOS.DSK in the current working directory, this file will automatically be loaded.

3.4.8
Desktop Save

This command saves the current desktop settings in the presently selected configuration file. The name of the current configuration file can be seen next to the command in the menu. If you wish to save the desktop settings under another name, use → Menu item Desktop Save As [Page 202]. In order to reload the saved desktop settings, use → Menu item Desktop Load [Page 202].

3.4.9
Desktop Save As

This command opens a → File dialog [Page 193] to save the desktop settings under a name to be specified. This new filename is then shown th the right of the → Menu item Desktop Save [Page 202].

3.5
Menu Windows

This menu offers commands to activate open windows and to modify them.

3.5.1
Next Window

This command activates the next window. If there is more than one window on the desktop, the one which is next in the internal list will be activated and put to the foreground. Alternatively, this can also be done with the right mouse button.

3.5.2
Previous Window

This command activates the previous window. If there is more than one window on the desktop, the previous window in the internal list will be activated and brought to the foreground.

3.5.3
Windowsize/-position

With this command you can begin to change the size and position of the active window: To change the size of the window use the arrow keys while holding down the shift key. To change the position of the window you just have to use the arrow keys. If you wish to keep the old position and size press the <Esc> key, otherwise confirm with the <Return> key.

3.5.4
Zoom Window

This command zooms the active window to its maximum size. A second execution of this command will decrease the window to its previous size.

3.5.5
Cascade

This command arranges the windows staggered one above the other increased by one line and one column.

3.5.6
Tile

This command tries to arrange the windows, so that no window is above another. Of course this will seldom work, because the sum of the windows is often higher than the size of the screen and some windows have a fixed size.

3.5.7
Window List

This command will open the →Dialog Window List [Page 194]. All titles of the open windows will be shown. Selecting one of these titles will activate the corresponding window.

3.6
Menu Help

Within the menu Help, you can gain access directly to main keywords of the online-help. From there you can navigate to any other theme.

3.6.1
Help Contents

Selecting this command you will get to the table of contents of the help system of CemoS.

3.6.2
Help Desktop

Selecting this command you will get help concerning the usage of the →Desktop [Page 186].

3.6.3
Tutorial

If you select this command, you will gain access to the →Tutorial [Page 183] of this program.

3.6.4
Help on the Active Model

If you select this command, you will receive specific help on the currently active model. You will get the same help if you select this model in the →Submenu Active Model [Page 199] and press <F1>.

3.6.5
Help Index

If you select this command, you will reach the alphabetically ordered keyword list from which you can jump directly to the corresponding help node.

3.6.6
Help About Help

If you select this command, you will get hints on the usage of the →Help System [Page 186].

3.6.7
Info

If you select this command, you will reach the →Information window [Page 195].

Chapter 4:
Description of the Model Interchange Format

The Model Interchange Format (MIF) which was developed for CemoS enables us to save all input parameters and model results in a comprehensible way. In particular the dependencies of parameters are considered as well as explanations and comments at diverse places. The MIF is a text format and hence easy to read and modify. Special parts are considered for the substance properties. Furthermore, the MIF is extendable and language independent, apart from the explanations and comments. Together with a concept for standard names the model output can be presented in many languages. A complete compiler exists for the MIF which is implemented as an awk-script (Aho, Kernighan, Weinberger, 1988).

CemoS is distributed with an awk-interpreter and some accompanied scripts which process the MIF files exported by CemoS.

The file STDNAMES.SN contains the translation into German and English of the standard names which are used by CemoS for the export of data.

CemoS only uses a small set of the MIF abilities to write complete model results and to read the substance and model parameters.

A more detailed description of the MIF is distributed together with CemoS.

4.1
Further Processing of CemoS MIF Files

With L_SELECT.BAT you can set the language to be used for the further processing of MIF files. There is a current and a standard language which are defined in a certain file. The comments you make on parameters, etc. will, of course, remain in the used language.

For the processing of a report from an MIF file REPORT.BAT or RE-PORTF.BAT can be executed. Both batch files need the specification of an MIF file as the first parameter. REPORT.BAT prints onto the screen, REPORTF.BAT needs a further parameter that specifies a filename on which to write the output.

A sample call could be: "REPORT test.mif", if you have already exported the file "test.mif". "REPORTF test.mif result" would write the report to the file "result". There is an equivalent command for the batch files described below that writes the output onto a file instead of onto the screen.

LIST.BAT lists all tables of an MIF file. TABLE.BAT and RAW.BAT extract the table data. TABLE.BAT produces an ouput with comments which can then be processed easily by gnuplot.

If you have installed gnuplot, you can view results as plots directly with the command PLOT.BAT which writes a postscript file of the plot via gnuplot. With the PLOT batch files you can also specify which columns to apply to each other. This is done by naming the numbers of the columns: First the column that represents the x-axis, then up to six columns for the y-axis.

If you have MIF files that do not look like CemoS MIF files, you have to use UREPORT.BAT to create a report of those data. This is needed if such MIF files contain aspects that CemoS could not export.

4.2
Creation of MIF Files for CemoS

The MIF files that are created by CemoS are almost self-explanatory. Within MIF files to be imported by CemoS you can leave out parameters or add your own ones. CemoS of course only imports the data known to it.

You can find more about this in the MIF documentation files.

4.3
Table of all Scripts for Further Processing

Command	Description
LIST mif	Shows the tables contained in a MIF file.
LISTF mif output	Writes the result of LIST into the file output.
L_SELECT [def] language	Sets the given language as the current one or (with "def") as the default language.
PLOT mif [X Y1 [...]]	The table found first in the MIF file "mif" will be presented on the screen via gnuplot which hence must be installed. X means the number of the column for the x-axis, Y1 to Y6 mean the column numbers for the y-axis. If X and Y are not defined, X = 1 is assumed and the remaining columns are taken as Y1, Y2, etc.

PLOTF mif [X Y1 [...]] output	Writes the plot that could be created with PLOT into the postscipt file "output".
TABLE [tab] mif	Shows the table with the name "tab" with all comments and with a space as column separator. Before comments the following character (#) is inserted. If "tab" is not specified, the table first found will be taken.
TABLEF [tab] mif output	Writes the output of TABLE into the file output.
RAW [tab] mif	Functions in a similar way to TABLE except that all comments will be droped.
RAWF [tab] mif output	Writes the output of RAW into the file "output".
REPORT mif	Creates a complete English report of the model simulation described in the MIF file "mif".
REPORTF mif output	Writes the output of REPORT into the file "output".
UREPORT mif	Creates (similarly to REPORT) a report from the data of an MIF file. But no special contents are assumed. Thus you can create reports of any MIF file.
UREPORTF mif output	Writes the output of UREPORT into the file.

Chapter 5:
Data Input

In this chapter we basically describe how to use the input dialogs and substance and scenario data bank.

For further information refer to:

→ Carrying out Simulation [Page 214],
→ Substance Data [Page 221].

For general information about the desktop elements see → Desktop [Page 186]. General information about using dialogs can be found under → General Dialog Elements [Page 189]

5.1
Input Substance Data

To enter a new substance use the menu-command → Menu item New Substance [Page 197]. If you wish to change or edit an existing substance use → Menu item Edit Substance [Page 197]. Executing one of these two commands will open the → Dialog Substance Data Input [Page 214] which enables you to input the substance data.

5.2
Input Model Data

To enter model parameters you have to execute the menu-command → Menu item Parameter input [Page 199]. If the active model has more than one parameter input dialog, the program opens a dialog with a list of these dialogs. Here you can select the input dialog you wish use.

You can find further details under → Model Parameter Input [Page 218].

5.3
Load Substance Data

You can store the substance data for further use in a substance data bank. The use of the substance data bank is explained in → Manage Substances [Page 215].

5.4
Load Scenarios

You can store the scenario data for further use in a scenario data bank. The use of the scenario-databank is explained in → Use of Scenarios [Page 219].

5.5
Import Data

CemoS provides the facility to import data from MIF files. The → Menu item Import [Page 196] was created for this purpose. If you select this item, a → File dialog [Page 193] is opened to select a file with data you wish to import. Then the → Dialog Contents of Import file [Page 210] appears in which all importable parts of the file are listed. After specifying the wanted data, it will be imported.

See also → Creation of CemoS MIF files [Page 207].

5.5.1
Dialog Contents of Import File

This dialog shows the various parts of a MIF file that can be imported in a selection list. Selecting an entry and confirming the dialog with the Ok button will start importing the wanted data. If you chose the Cancel button, the dialog is aborted and nothing is imported.

5.6
Estimate Input Data

Some of the values for substance data and model parameters can be estimated by using other (known) values. You can recognize whether an input value is estimable or not by the appearence of the estimate button at the bottom of the dialog. If you select the input line of an estimable value, the color of the estimate button turns from transparent to green. This means the estimate button is now selectable. Using the estimation button opens the → Dialog Estimation functions [Page 211]. Once you have selected an estimation function, the input line of the estimated value is displayed in another color. This indicates whether a value has been estimated or not. Overwriting an estimated value manually closes the use of the estimation function. This is indicated by the color of the corresponding input line changing.

5.6.1
Dialog Estimation Functions

This dialog is called up by the estimation button. It displays a list of the estimation functions that are implemented for the selected value. The name of the selected value is shown in the title of this dialog.

To use a certain estimation function from the list you have to select it and then press the Ok button or doubleclick on the selected estimation function. Using the Cancel button leaves the dialog without activating the estimation function.

For detailed information about the implementation of the estimation function you can use the Detail button. This opens the online-help with the corresponding help screen.

5.7
Comment Input Data

Most of the input data fields can be provided with a special comment. To comment upon a value you have to select the input line and click the button Comment. This opens the → Dialog Comment [Page 211].

5.7.1
Dialog Comment

This dialog is used to comment upon a single value. It contains a multi-line input field and the two buttons Ok and Cancel. If the value to be commented upon is estimated, the dialog contains, moreover, the name of the used estimation function. After you have edited the comment in the multi-line input field, you have to use the Ok button to accept the comment and to leave the dialog.

Chapter 6:
Data Output

6.1
Output to Screen

The screen output of simulation results is divided thematically into a certain number of windows. In addition to the regular results some models offer a → Table window [Page 189] and one or more simple graphic windows which present the tabular data graphically.

Besides the text graphics CemoS is also able to present high resolution graphics. You can get this presentation with → Menu item Show Graphics [Page 200].

6.2
Dialog Choose Graphic

This dialog offers a list of graphics that belong to the current model results. To view a graphic mark the corresponding item and click the button Ok or doubleclick the item directly.

6.3
Export Data

CemoS saves all simulation data in → Model Interchange Format [Page 206] via → Menu item Export As [Page 196]. To select this menu command a → File dialog [Page 193] appears where you specify the file which CemoS should export to.

These result files can be processed in many ways; see → Further processing of CemoS MIF files [Page 206].

Any further processing of CemoS simulation results should be done via this mechanism, because comprehensibility is then guaranteed.

6.4
Output to Printer

There is no direct connection from CemoS to a printer. In order to print you have to leave CemoS or open a DOS prompt. Before doing that you have to export the desired data the →Menu item Export As [Page 196].

If all the available information has been stored in an MIF file, the desired information can then be filtered by prepared or your own batch scripts.

For example the batch file REPORTF.BAT creates a nice English report of all data that an MIF file contains and writes it onto another file. This resulting file can be sent to the printer (or any other ouput device) with common DOS commands.

Further information can be found under →Further processing of CemoS MIF files [Page 206].

6.5
Output to File

The output of data into a file is the natural step before output to a printer. You can find the relevant information under →Output to Printer [Page 213].

6.6
Output of Tables

While exporting, result tables are written with all other available information into an MIF file. From this file you can extract the desired tables with prepared or your own batch scripts and put them into the needed shape.

The prepared batch scripts are LIST.BAT, RAW.BAT and PLOT.BAT. Further information can be found under →Further processing of CemoS MIF files [Page 206].

Chapter 7:
Carrying out Simulation

7.1
Substance Data Input

Use the →Menu item New Substance [Page 197] command to enter a new substance and use →Menu item Edit Substance [Page 197] if you wish to change the data of an existing substance. These two commands open the →Dialog Substance Data Input [Page 214] so the substance data can be entered.

7.1.1
Dialog Substance Data Input

This dialog is used to enter the physico-chemical data of a substance. It is opened by →Menu item New Substance [Page 197] if you wish to enter a new substance, or →Menu item Edit Substance [Page 197] if you only want to change the current substance data.

You can enter a value in two ways. First you can type the value in the input line manually. The second way is to use an estimation function. You can see whether an estimation function is available or not by the highlighted estimate button. Details about using estimation functions can be found under →Estimate Input Data [Page 210].

Furthermore, you can enter a comment for each value. To do so you have to select the input line you wish to comment upon and then press the comment button at the bottom of the dialog. Whether a value can be commented on or not can be seen by the color of the comment button. If the button appears in the same color as the dialog background, the linked value cannot be provided with a comment. More details about comments can be found under →Comment Input Data [Page 211].

Below the input lines the dialog contains an input field. This input field contains an overall comment for the whole substance data.

7.2
Substance Data Output

The program's output of a substance is a list of all substance data that were entered in the →Dialog Substance Data Input [Page 214].

Details about the output procedures of CemoS can be found under:

→Output to Printer [Page 213],
→Output to File [Page 213].

7.3
Manage Substances

In order to only have to enter the physico-chemical data of a substance of interest once you can store this information in a databank from which it can be retrieved for further use. The following commands are available here:

7.3.1
Load Substance

You can load a substance from the open substance database by using the →Menu item Load Substance [Page 198]. It opens the →Dialog Load Substance [Page 215] that allows you to select the substance you wish to load.

In order to load a substance you first have to open a database with at least one substance stored. If you have not opened a database or if the database does not contain substances, this menu item is not selectable. The default database is called SUBDATA.DAB and is open unless you open another one.

Dialog Load Substance

This dialog enables you to select and load a substance from the open database.

It contains a list with two columns. This list contains the name and the CAS-number of each of the stored substances. A missing CAS-number is indicated by an 'unknown'. The list is supplied with a horizontal scrollbar, because the substance name can be wider than the list. Using the scrollbar shows the rest of the name.

You can search for a substance in the list by pressing the Search button. It opens the →Dialog Search Substance [Page 216] that allows you to enter the search text. A search always starts at the top of the list and does not depend on the selected entry.

If a match is found, the matching item is selected. Furthermore, the Continue button becomes selectable. This button can be used to continue the search with the same text. A continued search starts one entry after the last hit. As long as the search is successful, the Continue button remains selectable. If the search does not find any more matching items, a message box is displayed and the Continue button is no longer unselectable. For a new search you have to use the Search button again.

In order to load a substance you have to select the item in the list and press the Load button or doubleclick on the item.

Dialog Search Substance

This dialog enables you to enter two search terms (the substance name and the CAS-number). These terms are used to search for a substance in the substance list of the → Dialog Load Substance [Page 215] or the → Dialog Delete Substance [Page 217].

The implemented search routine uses the following rules:

• you can only enter one search term
• if the CAS-number is unknown, you do not have to enter 'unknown', you can leave the input line empty
• all items which contains the search terms will be found
• if you have entered a search term in both input lines, the search will only find the items where both terms match (logical and).

The search starts by pressing the Ok button and the first matching item will be selected. If you continue the search by pressing the Search button, the old search term will be used again.

7.3.2
Save Substance

You can save a substance in the open substance database by using → Menu item Save Substance [Page 198]. This enables you to retrieve the substance data from the database at any time.

In order to save a substance you have to open an existing database (→ Open Database [Page 217]) or create a → New Database [Page 217]. This menu item is not selectable if no databases are open. The default database is called SUB-DATA.DAB and is open unless you open another one.

7.3.3
Delete Substance

You can delete a substance from a substance database by using the → Menu item Delete Substance [Page 198]. It opens the → Dialog Delete Substance [Page 217], where you can select the substance you wish to delete.

To delete a substance you have to open a database with at least one substance stored. If you haven't opened a database or if the database does not contain any substances, this menu item is not selectable.

Dialog Delete Substance

This dialog is very similarly structured to the →Dialog Load Substance [Page 215]. The difference is that it is indicated by a Delete button rather than a Load button.

Before the program deletes a substance you will be asked if you really want to delete the selected substance. This question prevents the unwanted deletion of a substance by a doubleclick.

7.3.4
New Database

You can create a new substance database by using the →Menu item New Database [Page 198]. The new database automatically becomes the currently open one. It does not contain any substances. In order to store new substances you have to enter them using →Menu item New Substance [Page 197] command and then store them into the new database with →Menu item Save Substance [Page 198].

The new database will be created in the current working directory. If you wish to create the database in another directory you can enter the path of the directory in the file dialog New Database or you can change the current directory with →Menu item Change Dir [Page 197] before creating the database.

7.3.5
Open Database

You can open an existing substance database by using →Menu item Open Database [Page 198]. It opens a →File dialog [Page 193] to select or enter the name of the database. To create a new database you have to use →Menu item New Database [Page 198].

If the database is not in the current directory, you can change the directory with →Menu item Change Dir [Page 197] before opening the database.

7.3.6
Merge Databases

You can merge the current database with a second database by using →Menu item Merge Databases [Page 198]. Therefore a →File dialog [Page 193] is opened that allows you to select the second database.

After using the Ok button of the file dialog, the substances in the second database will be copied into the current database. If an identical substance name is found in both databases, the program asks you whether the substance of the current database should be overwritten by the substance in the second database or not. If you no longer know which substance is the correct one, you should select No, to prevent the substance in the current database from being deleted. If you select No, neither of the substances will be deleted. They both remain in their databases.

The second database has to carry a different name to the current database.

7.3.7
Delete Database

You can delete a substance database by using → Menu item Delete Database [Page 198]. This will open a → File dialog [Page 193] to select the database you want to delete.

7.4
Select Models

In order to select the current model you have to use → Submenu Active Model [Page 199]. If you select the corresponding menu item, the model will be activated. Selecting a new model causes the closure of any remaining result windows at the desktop; even if you select the old model the desktop will be reset. The active model is marked in the menu with a tick. Furthermore, the name of the active model is displayed in the → Status window [Page 188].

7.5
Model Parameter Input

In order to enter the model parameters you have to use → Menu item Parameter input [Page 199]. If the model consists of more than one dialog, the program opens the → Dialog Parameter Groups [Page 218]. This dialog allows you to select the input dialog where the parameters you want to change can be found. After your selection the program opens all input dialogs you have selected in sequence.

7.5.1
Dialog Parameter Groups

This dialog contains a list of all input dialogs that exist for the active model. The set of parameters was split for some models. This means that the input values are arranged in several dialogs. This was done for clarity.

You can choose one or more item from this list to open the dialogs you need. To select all items within one step you can press the All button.

Once you have selected the desired items, you can close the dialog by pressing the Ok button and the program will open all selected input dialogs.

7.6
Model Parameter Output

The program's output of the model parameter is a list of all used parameters. This means that for some models (e.g. Plume), only the active settings are printed out.

Details about the output can be found under:
→ Output to Printer [Page 213],
→ Output to File [Page 213].

7.7
Use of Scenarios

For several reasons it is important to store the complete parameter set of a simulation run. CemoS provides this facility with its scenarios. Scenarios allow you or someone else to rerun a simulation with exactly the same set of values or to just vary some important values for sensitivity analysis.

Every model has its own scenario file in which you can save different runs of the same model. A scenario contains the model parameters and the substance data. The commands that are used to manage scenarios are described in the following section. All commands open the → Dialog Scenarios [Page 219].

7.7.1
Dialog Scenarios

This dialog is used to load, save and delete scenarios.

It contains two input lines. The first one contains the name of the scenario. The second contains a short description of it. It is not necessary to enter the name of the model in the description, because the scenarios are sorted by models, as we will see below.

Below the input lines there is a list of all previously saved scenarios. The list consists of three columns. The first one shows all scenario files in the current scenario directory. The second list states the names of the models to which the scenarios belong. The latter list shows a short description of each stored scenario. If you select a disk drive or a subdirectory in the first column that does not contain any scenario files, the second column informs you whether it is a directory or a disk drive. If you wish to load, delete or overwrite a scenario, you can select it from this list with a double-click.

7.7.2
Load Scenario

You can load a scenario by using →Menu item Scenario Load [Page 199]. It opens the →Dialog Scenarios [Page 219] to select the scenario you wish to load.

7.7.3
Save Scenario

You can save the current parameter set by using →Menu item Scenario Save [Page 200]. If a scenario file has been loaded or saved before, the current parameter set will be saved with the current scenario name. If no current (saved) scenario exists, this command has the same effect as →Save Scenario as [Page 220].

7.7.4
Save Scenario as

You can save the current parameter set with a new name by using →Menu item Scenario Save as [Page 200]. It opens the →Dialog Scenarios [Page 219] to input the new name and to save the parameter set under this name.

7.7.5
Delete Scenario

You can delete a scenario in the scenario file by using →Menu item Scenario Delete [Page 200]. It opens the →Dialog Scenarios [Page 219] to select the scenario you wish to delete.

7.7.6
Scenario Directory

The default scenario directory that is shown when opening the scenario dialog is the SCENARIO subdirectory of the CemoS program directory. If this subdirectory is not existing CemoS searchs in the program directory. If you want CemoS to search in another directory you can specify this the →Dialog Change Directory [Page 193]. This command opens the →Menu item Scenario Directory [Page 200] where you change to the wanted directory.

Chapter 8:
Substance Data

Physico-chemical data exist for each substance. Each value can be provided with a comment. If a value is estimated, the applied estimation method can be seen in the comment window.

CemoS does not just accept any values. Warnings will be prompted if values are outside of reasonable ranges. It is not possible to enter physically impossible values. If this occurs, CemoS will prompt an error message.

Description of the substance data:

Notation	Unit	Warning range	Error range
Name	–	–	–

Name or other specification for this substance.

CAS	–	–	–

CAS-Number of the substance; CAS means Chemical Abstract Service. The syntax for such a number is: Many, but at least one digit followed by a hyphen followed by two digits followed by a hyphen followed by one digit ([...Z]Z-ZZ-Z).

Sum formula	–	–	–

Sum formula of the substance. If the molar mass is to be estimated from this formula, a correct input is neccessary. For the syntax see → Estimation of molar mass [Page 260].

M	g/mol	[1; 1000]	[1; ∞[

Molar mass of the substance. See also → Estimation of the molar mass [Page 260].

bp	K	[mp; 5000]	[mp; ∞[

Boiling point of the substance. (Will be compared with the mp when the input dialog has been left).

mp	K	[0; 5000]	[0; ∞[

Melting point of the substance.

Notation	Unit	Warning range	Error range
K_{AW}	–	[0; 100]	[0; ∞[

Partition coefficient air/water (dimensionless Henry constant). Estimation: See → Estimation of K_{AW} [Page 261].

K_{OC}	cm^3/g	[0; $1 \cdot 10^9$]	[0; ∞[

Partition coefficient organic carbon/water. This value describes the sorption of lipophile organic matter to organic substances (humus, etc.) Estimation: See → K_{OC} Estimations [Page 262].

log K_{OW}	–	]−∞; 15]	]−∞; 15]

Logarithm of the partition coefficient octanol/water. This value describes the lipophility of the substance. Estimation: See → log K_{OW} Estimations [Page 263].

BCF	–	[1; $1 \cdot 10^{15}$]	[1; $1 \cdot 10^{15}$]

Bioconcentration factor (concentration in fish / concentration in surrounding water). Estimation: See → BCF Estimations [Page 264].

Deg_{Water}	1/d	[0; 36]	[0; 36]

First-order degradation rate in water.

Deg_{Soil}	1/d	[0; 36]	[0; 36]

First-order degradation rate in soil.

Deg_{Plant}	1/d	[0; 36]	[0; 36]

First-order degradation rate in plants.

Deg_{Air}	1/d	[0; 36]	[0; 36]

First-order degradation rate in the air.

D_a	m^2/d	]0; 10]	]0; 10]

Diffusion coefficient of the substance in gas. Estimation: See → Estimation for diffusion coefficient air [Page 264].

D_w	m^2/d	]0; $1 \cdot 10^{-3}$]	]0; $1 \cdot 10^{-3}$]

Diffusion coefficient of the substance in water. Estimation: See → Estimations of diffusion coefficient water [Page 265].

VP	Pa	[0; 110000]	[0; ∞[

Vapor pressure of the substance at 20°C.

pKa	–	[0; 14]	[−10; 24]

The *pKa* is the negative decadic logarithm of the dissociation constant.

ws	g/l	[0; 500]	[0; ∞[

Water solubility. Estimation: See → Water solubility estimation [Page 265].

Notation	Unit	Warning range	Error range

Protolysis – – –

Protolysis of the substance. This could include the acidic, alkaline and neutral states. With this information some models could execute a $\rightarrow K_{AW}$-Standard correction [Page 275].

Comment – – –

General comment on the substance. This should cover information about the origin of the data and other important hints. It is recommended here to add the date of the substance input and the name of the person who entered the data.

Chapter 9:
Models

In this chapter the technical aspects of the implemented models are explained. These technical references should not be understood as general descriptions of the models. Instead, all the compuations undertaken by CemoS are presented and hence become comprehensible.

How to run a model is described in →Carrying out Simulation [Page 214].

A description of a model consists of several parts: First a (very) short general description is given followed by a table of the output parameters with units and the notation which is used in the equations. After that the complete model computation followed by a table of the helping variables is presented. Then the used substance data and any error messages that could appear are listed before the table of input parameters ends the model description. This table contains the notations that are used in the equations, units, default values, warning and error ranges and other hints. The hints about the parameters that are not needed by the model can also be found there. These are mainly the parameters which are added to estimate necessary and possibly unknown data.

9.1
Air

This model calculates the atmospheric transport of aerial emissions assuming constant atmospheric and meteorological properties.

The output is divided into two groups: general results and a mass balance:

Description	Unit	Notation
Concentration in the box	kg/m^3	C_0
Total deposition	$kg/(m^2 \cdot a)$	t_{dep}
Annual inhalation dose	kg/a	$inhal_{dose}$
Dry particle deposition	kg/a	dry_{par}
Wet particle deposition	kg/a	wet_{par}
Dry gaseous deposition	kg/a	dry_{gas}

Description	Unit	Notation
Wet gaseous deposition	kg/a	wet_{gas}
Photo degradation	kg/a	$photo_{deg}$
Advection	kg/a	$advection$

The model carries out the following calculations:

$$area = x \cdot y$$

$$volume = area \cdot z$$

$$v_{dep,dry} = (1 - f_p) \cdot VdGasDry + f_p \cdot VdParDry$$

$$v_{dep,wet} = \frac{\left(\frac{1-f_p}{K_{AW}} + f_p \cdot washoutRatio\right) \cdot rain}{8.64 \cdot 10^7}$$

$$v_{dep} = v_{dep,dry} + v_{dep,wet}$$

$$C_0 = \frac{input}{v_{Wind} \cdot 86400 \cdot y \cdot z + v_{dep} \cdot area \cdot 86400 + Deg_{Air} \cdot volume}$$

$$t_{dep} = v_{dep} \cdot C_0 \cdot 365 \cdot 86400$$

$$inhal_{dose} = inhal \cdot C_0 \cdot 365$$

$$dry_{par} = f_p \cdot VdParDry \cdot C_0 \cdot area \cdot 86400 \cdot 365$$

$$wet_{par} = \frac{f_p \cdot washoutRatio \cdot rain \cdot C_0 \cdot area \cdot 86400 \cdot 365}{8.64 \cdot 10^7}$$

$$dry_{gas} = (1 - f_p) \cdot VdGasDry \cdot C_0 \cdot area \cdot 86400 \cdot 365$$

$$wet_{gas} = \frac{(1 - f_p) \cdot rain \cdot C_0 \cdot area \cdot 86400 \cdot 365}{K_{AW} \cdot 8.64 \cdot 10^7}$$

$$photo_{deg} = Deg_{Air} \cdot volume \cdot C_0 \cdot 365$$

$$advection = v_{Wind} \cdot 86400 \cdot y \cdot z \cdot C_0 \cdot 365$$

with

Notation	Description	Unit
$washoutRatio$	Washout efficiency of precipitation (constant $= 2 \cdot 10^5$)	$(kg/m^3 \ rain)/$ $(kg/m^3 \ air)$
$area$	Area of box	m^2
$volume$	Volume of box	m^3
$v_{dep,dry}$	Total dry deposition velocity	m/s
$v_{dep,wet}$	Total wet deposition velocity	m/s
v_{dep}	Total deposition velocity	m/s

K_{AW} and Deg_{Air} are → Substance Data [Page 221].

Error messages:

Condition	Reason
K_{AW} of substance is 0.	The model is not applicable for a substance with $K_{AW} = 0$.

Description of the input parameter:

Notation	Unit	Default	Warning range	Error range
inhal Inhalation rate.	m^3/d	20	[2; 40]	$[0; \infty[$
input Release rate of substance.	kg/d	1	$[0; \infty[$	$[0; \infty[$
f_p Particle fraction. Estimation: See $\rightarrow$ Estimation of particle fraction [Page 270].	–	*unknown*	[0; 1]	[0; 1]
rain Precipitation intensity.	mm/d	2.1	[0; 2000]	$[0; \infty[$
v_{Wind} Average vertical wind velocity.	m/s	1	[0; 20]	$[0; 3 \cdot 10^8]$
VdGasDry Dry gaseous deposition velocity. Estimation: See $\rightarrow$ VdGas-Estimation [Page 266].	m/s	*unknown*	[0; 1]	$[0; 3 \cdot 10^8]$
VdParDry Dry particle deposition velocity.	m/s	0.01	[0; 1]	$[0; 3 \cdot 10^8]$
x Length of box in main direction of wind (maximum is circumference of the earth).	m	1000	$[10; 100 \cdot 10^3]$	$[1; 40 \cdot 10^6]$
y Width of the box.	m	1000	$[10; 100 \cdot 10^3]$	$[1; 40 \cdot 10^6]$
z Height of the box.	m	500	$[10; 10 \cdot 10^3]$	$[1; 20 \cdot 10^3]$

9.2
Buckets

Buckets is a discrete cascade model for the simulation of the advective transport of pollutants in soil. In contrast to all other implemented models, this model is based on a numerical method (Finite-Differences Method). The time step is fixed and is equal to one day. Due to this approach, an error (numerical dispersion) is unavoidable. Buckets allows the simulation of inhomogeneous layers and of non-stationary water conditions. This model cannot be used for compounds with high volatility (high K_{AW}). The fundamental equations of the model is the tipping buckets approach by Burns which was derived by Richter (1990). The equations were extended.

The user is able to simulate with or without the transport of solutes in the capillary rise. The parameters water content, concentration of solutes, precipitation, evaporation and runoff have to be inserted explicitly for each layer and time step, respectively. The parameters field capacity and wilting point are identical for each layer.

The output consists of the solute concentration and water content of each layer. Furthermore, fractions of the chemical in aqueous phase and matrix are also given. In addition, the total leached amount of water and solutes and the total amount of evaporated water is calculated. The maximal Courant number and numerical dispersion are also calculated.

Description	Unit	Notation
Solute concentration	kg/m^3	$C[2]$
Water content	m^3/m^3	$V[2]$
Fraction in aqueous phase	–	f_W
Fraction in soil matrix	–	f_M
Amount of leaching:		
water content	m^3/m^2	$WOutI$
solute content	kg/m^2	$SOutI$
Amount of evaporation:		
water content	m^3/m^2	$WOutE$
Courant number	–	CR
Numerical dispersion	m^2/d	D_{Num}

At the beginning the $\rightarrow K_{AW}$ standard correction will automatically be carried out [Page 275] . If K_{AW} is acceptable (condition: $K_{AW} < 10^3$), the following computations will be carried out:

(0) $WOutI = SOutI = WOutE = uMax = 0$
(1) $Number\ of\ layers = DepthTo\ /\ d$

(2) $OM = 1.724 \cdot OC$
(3) $Density = (1 - Porosity) \cdot (OM \cdot 1400 + (1 - OM) \cdot 2650)$
(4) $K_{MW} = K_d \cdot Density \cdot 10^{-3}$

(5) For all layers:
$$f_W = V[1]/(K_{MW} + V[1])$$
(6) For all layers:
$$W[1] = V[1] \cdot d$$
$$S[1] = C[1] \cdot W[1] \cdot (f_W/V[1])$$

$water\ balance = (precipitation - evaporation - runoff) \cdot 10^{-3}$

IF $|waterbalance|/(K_{MW} + wiltingpoint) > uMax$ THEN
$\quad uMax = |waterbalance|/(K_{MW} + wiltingpoint)$
$\quad CR = uMax/d$
$\quad$ IF $CR > 1$ THEN abort simulation

IF $waterbalance < 0$
THEN
$\quad$ Process of evaporation
ELSE
$\quad$ Process of infiltration and leaching

For all layers:
$$C[2] = S[2]/(W[2] \cdot (f_W/V[1]))$$
$$V[2] = W[2]/d$$
$$f_W = \frac{V[2]}{K_{MW}+V[2]}$$
$$f_M = \frac{K_{MW}}{K_{MW}+V[2]}$$
$$V[1] = V[2]$$
$$C[1] = C[2]$$

$WOutI = WOutI + leached\ amount\ of\ water$
$SOutI = SOutI + leached\ amount\ of\ solutes$
$WOutE = WOutE + evaporated\ amount\ of\ water$

(7) Repeat (6) for all time steps
(8) $D_{Num} = 0.5 \cdot (uMax \cdot d - uMax^2)$

with

Notation	Description	Unit
K_{MW}	distribution coefficient matrix/water	–
OM	content of organic carbon	–
$Density$	density of soil	kg/m^3
$W[1]$	old content of water	m^3/m^2
$W[2]$	new content of water	m^3/m^2
$S[1]$	old content of solutes	kg/m^2
$S[2]$	new content of solutes	kg/m^2
$WPlus$	infiltrated amount of water	m^3/m^2
$SPlus, SOut$	infiltrated amount of solutes	kg/m^2
$WMax$	max. possible infiltrated, evaporated content of water	m^3/m^2
$WMinus$	amount of evaporated water	m^3/m^2
$WEff$	effective amount of evaporated water	m^3/m^2
$SMinus$	amount of solutes in capillary rise	kg/m^2
$u\,Max$	Maximal flow velocity	m/d
t	time step	d

Other notations are $\rightarrow$ Substance Data [Page 221].

Error messages:

Condition	Reason
K_{AW} of substance is $\geq 10^{-3}$.	This model is not suitable for chemicals which are transported in gaseous phase primary.
Courant number is more than 1.	It is not possible to simulate using the current parameters. Increase thickness of layers.
Water content is more than field capacity.	It is not possible to start simulation if water content is more than field capacity.
Water content is less than wilting-point.	It is not possible to start simulation if water content is less than wilting point.

Process of infiltration and leaching

The process of infiltration and leaching will be used if the water balance is positive.

(1) $WPlus = waterbalance$
(2) $SPlus = input \cdot WPlus$
(3) $WMax = W[1] + WPlus$

(4) $WPlus = WMax - fieldcapacity \cdot d$

(5) IF $WPlus > 0$ THEN

   $SOut = S[1] \cdot WPlus/WMax$

   ELSE

   $SOut = 0$

   $S[2] = S[1] \cdot e^{-Deg_{soil} \cdot t} + SPlus \cdot f_W - SOut \cdot f_W$

(6) IF $WPlus \geq 0$ THEN

   $W[2] = fieldcapacity \cdot d$

   $SPlus = SOut$

   Repeat (3) to (6) for all other layers

   ELSE

   $W[2] = WMax$

   For all following layers:

   $W[2] = W[1]$

   $S[2] = S[1] \cdot e^{-Deg_{soil} \cdot t}$

Process of Evaporation

The process of evaporation will be used if the water balance is negative.

(1) $WMinus = |waterbalance|$

(2) $WMax = W[1] - wiltingpoint \cdot d$

   IF $WMax < 0$

   THEN

   $WMax = 0$

(3) IF $WMax > WMinus$

   THEN

   $WEff = WMinus$

   ELSE

   $WEff = WMax$

(4) $W[2] = W[1] - WEff$

(5) For the first layer:

   $S[2] = S[1] \cdot e^{-Deg_{soil} \cdot t}$

   For all other layers:

   IF $capillary\ rise$

   THEN

   $SMinus = S[1] \cdot WEff/W[1]$

   $S[2] = S[1] \cdot e^{-Deg_{soil} \cdot t} - SMinus \cdot f_W$

   Process of capillary rise

   ELSE

   $S[2] = S[1] \cdot e^{-Deg_{soil} \cdot t}$

(6) IF $WMax \geq WMinus$

   THEN

For all remaining layers
$W[2] = W[1]$
$S[2] = S[1] \cdot e^{-Deg_{Soil} \cdot t}$
end of process
ELSE
$WMinus = WMinus - WMax$
(7) Repeat (3) to (6) for all remaining layers

Process of capillary rise

The process of capillary rise is part of the process of evaporation. This process begins at the layer above the current one.

(1) $SOut = S[1] \cdot WEff/(W[1] + WEff)$
(2) $S[2] = S[2] \cdot e^{-Deg_{Soil} \cdot t} + SMinus \cdot f_W - SOut \cdot f_W$
(3) $SMinus = SOut$
(4) Repeat (1) to (3) for all remaining layers between the current one and the soil surface.

Description of input parameter:

Notation	Unit	Default	Warning range	Error range
d	m	0.1	[0.01; 50]	[0.01; 6300000]
Thickness of a layer.				
Wiltingpoint	m^3/m^3	0.05	[0; 1]	[0; 1]
Minimal concentration of water for evaporation.				
Fieldcapacity	m^3/m^3	0.35	[0; 1]	[0; 1]
Maximal water content of a layer.				
Input	kg/m^3	0.1	[0; 500]	[0; 1300]
Concentration of substance in infiltrating precipitation.				
Time steps	d	20	[1; ∞[	[1; ∞[
Time steps of calculation. One time step stands for one day. The maximum of time steps depends on available memory space.				
Porosity	m^3/m^3	0.5	]0; 1]	]0; 1]
Porosity of soil.				
K_d	cm^3/g	*unknown*	[0; 1 · 10^6]	[0; ∞[
Distribution coefficient matrix/water. Estimation: See → K_d-Estimation [Page 266].				

OC	kg/kg	0.02	[0; 1]	[0; 1]

Content of organic carbon.

pH	–	6	[2; 12]	[0; 14]

pH of soil (only necessary to check the corrected K_{AW}).

DepthFrom	m	0	[0; 50]	[0; 6300000]

Depth of soil at which computation should begin.

DepthTo	m	2	[0.01; 50]	[0.01; 6300000]

Depth of soil at which computation should end.

Capillary rise	–	*on*	–	–

Switch which determines if capillary rise will be computed with or without transport of solutes.

V[1]	m^3/m^3	0.3	]0; 1]	]0; 1]

Volumetric water content of a layer before computation.

C[1]	kg/m^3	0	[0; 500]	[0; 1300]

Solute concentration of a layer before computation.

Precipitation	mm/d	2.1	[0; 2000]	[0; ∞[

Amount of precipitation in each time step.

Evaporation	mm/d	1.6	[0; 1999]	[0; ∞[

Amount of evaporation in each time step.

Runoff	mm/d	0.2	[0; 1999]	[0; ∞[

Runoff on the surface of soil in each time step.

9.3
Chain

This food chain model consists of three parts called the producer, consumer 1 and consumer 2. Actually this model simulates a general cascade with three elements. Therefore nuclear decay could also be simulated.

The output consists of nine value groups, time/substance mass and time/substance concentration diagrams for producer and consumers, these diagram data as a table, the remaining concentrations and a mass balance:

Description	Unit	Notation
Substance mass in producer	kg	m_P
Substance mass in consumer 1	kg	m_{C1}
Substance mass in consumer 2	kg	m_{C2}
Substance concentration in producer	kg/kg	$conc_P$
Substance concentration in consumer 1	kg/kg	$conc_{C1}$
Substance concentration in consumer 2	kg/kg	$conc_{C2}$

The mass balance is also given in percentage of total mass.

The model carries out the following calculations:

$$m_P = input \cdot e^{-(k_{12}+degr_P) \cdot t_{end}}$$

$$m_{C1} = \frac{input \cdot k_{12} \cdot (e^{-(k_{12}+degr_P) \cdot t_{end}} - e^{-(k_{23}+degr_{C1}) \cdot t_{end}})}{(degr_{C1} + k_{23} - degr_P - k_{12})}$$

$$m_{C2} = input \cdot k_{12} \cdot k_{23}$$
$$\cdot \left(\frac{e^{-(k_{12}+degr_P) \cdot t_{end}}}{(degr_{C1} + k_{23} - degr_P - k_{12})(degr_{C2} - degr_P - k_{12})} \right.$$
$$+ \frac{e^{-(k_{23}+degr_{C1}) \cdot t_{end}}}{(degr_{C1} + k_{23} - degr_P - k_{12})(degr_{C1} + k_{23} - degr_{C2})}$$
$$\left. + \frac{e^{-degr_{C2} \cdot t_{end}}}{(degr_{C2} - degr_{C1} - k_{23})(degr_{C2} - degr_P - k_{12})} \right)$$

$$m_{degr} = input - m_P - m_{C1} - m_{C2}$$

$$conc_P = \frac{m_P}{mass_P}$$

$$conc_{C1} = \frac{m_{C1}}{mass_{C1}}$$

$$conc_{C2} = \frac{m_{C2}}{mass_{C2}}$$

Error messages:

Condition	Reason
$k_{12} + degr_P = k_{23} + degr_{C1}$ or $k_{12} + degr_P = degr_{C2}$ or $k_{23} + degr_{C1} = degr_{C2}$	The above equations are only valid for different egenvalues in pairs. For eigenvalues with multiples 2 or 3, solutions exist but they are not considered in this version of CemoS.

No substance data are used.

Description of input parameter:

Notation	Unit	Default	Warning range	Error range
mass$_P$	kg	$1 \cdot 10^4$	$[1 \cdot 10^{-8}; \infty[$	$[1 \cdot 10^{-10}; \infty[$
Mass of producer.				
mass$_{C1}$	kg	500	$[1 \cdot 10^{-9}; \infty[$	$[1 \cdot 10^{-11}; \infty[$
Mass of consumer 1.				
mass$_{C2}$	kg	25	$[1 \cdot 10^{-10}; \infty[$	$[0; \infty[$
Mass of consumer 2.				
degr$_P$	1/d	0.05	$[0; \infty[$	$[0; \infty[$
Degradation rate of substance in producer.				
degr$_{C1}$	1/d	0	$[0; \infty[$	$[0; \infty[$
Degradation rate of substance in consumer 1.				
degr$_{C2}$	1/d	1	$[0; \infty[$	$[0; \infty[$
Degradation rate of substance in consumer 2.				
k$_{12}$	1/d	0.0035	$[0; \infty[$	$[0; \infty[$
Transfer rate from producer to consumer 1.				
k$_{23}$	1/d	0.017	$[0; \infty[$	$[0; \infty[$
Transfer rate from consumer 1 to consumer 2.				
input	kg	$100 \cdot 10^{-6}$	$[0; \infty[$	$[0; \infty[$
Input of substance to producer.				
t$_{end}$	d	100	$[1; 365 \cdot 10^4]$	$[0; \infty[$
Duration of simulation.				
t$_{step,num}$	–	100	$[10; 200]$	$[1; 2000]$
Number of output intervals.				

9.4
Level 1

The model Level 1 is implemented by following the fugacity approach of Donald Mackay (Mackay, 1991). Instead of fugacities, partition coefficients are used. Level 1 computes the equilibrium distribution of a chemical in the compartments water, air, soil, sediment, suspended matter, fish and plant. An instantaneous partition equilibrium is assumed. The following results are computed:

Description	Unit	Notation
Concentration of the chemical in each compartment	kg/m^3	c[1...7]
Mass of the chemical in each compartment	kg	m[1...7]

The input parameters are described at the end of the section →Level 2 [Page 237].

The model consists of the following equations:

$i = 0$
$Kcorr = 1$
IF compartment Water selected THEN
    $i = i + 1$
    $K[i] = 1$
    $V[i] = Water.Area \cdot Water.Depth$
END

IF compartment Air selected THEN
    $i = i + 1$
    IF $i > 1$ THEN
      $K[i] = K_{AW}$
    ELSE
      $K[i] = 1$
      $Kcorr = K_{AW}$
    END
    $V[i] = 0$
    IF compartment Water selected THEN
      $V[i] = Water.Area$
    END
    IF compartment Soil selected THEN
      $V[i] = V[i] + Soil.Area$
    END
    $V[i] = V[i] \cdot Air.Height$
END

IF compartment Soil selected THEN
    $i = i + 1$
    $K[i] = Soil.K_{BW}/Kcorr$
    $V[i] = Soil.Area \cdot Soil.Depth$
END

IF compartment Sediment selected THEN
    $i = i + 1$
    $K[i] = Sediment.K_{SW}/Kcorr$
    $V[i] = Sediment.Area \cdot Sediment.Depth$
END

IF compartment Suspended matter selected THEN
    $i = i + 1$

$$K[i] = Suspend.K_d \cdot (Suspend.rho/1000)/Kcorr$$
$$V[i] = \frac{(Water.Depth \cdot Water.Area \cdot Suspend.Content)}{Suspend.rho}$$
END

IF compartment Fish selected THEN
  $i = i + 1$
  $$K[i] = BCF \cdot (Fish.rho/1000)/Kcorr$$
  $$V[i] = \frac{Water.Depth \cdot Water.Area \cdot Fish.Content}{Fish.rho}$$
END

IF compartment Plant selected THEN
  $i = i + 1$
  $$K[i] = Plant.K_{PW}/Kcorr$$
  $$V[i] = \frac{Soil.Area \cdot Plant.Vegetation}{Plant.rho}$$
END

$$max = i$$

$$c[1] = \frac{Gen.Input}{V[1] + K[2] \cdot V[2] + \ldots + K[max] \cdot V[max]}$$
$$m[1] = V[1] \cdot c[1]$$

FOR $i = 2$ TO max DO
  $c[i] = K[i] \cdot c[1]$
  $m[i] = V[i] \cdot c[i]$
END

with

Notation	Description	Unit
$K[1 \ldots 7]$	Distribution coefficient ref. Water or Air	–
$V[1 \ldots 7]$	Volume of each compartment	m³
$Kcorr$	correction coefficient	–
max	Number of selected compartments	–
i	Dummies	–

For the values K_{AW}, BCF see → Substance Data [Page 221].

9.5
Level 2

The model Level 2 is implemented by following the fugacity approach of Donald Mackay (Mackay, 1991). Instead of fugacities, distribution coefficients are used. Level 2 computes the equilibrium distribution of a chemical in the compartments water, air, soil, sediment, suspended matter, fish and plant for a user defined period of time and for the steady-state solution. Decomposition rates and thru-flux quantities can be entered for each compartment.

The output of a simulation run consists of the concentration, the masses and the mobility of the substance for each compartment. Furthermore, the mass balance is shown for steady-state and after the chosen time period. The following results are also computed:

Description	Unit	Notation
concentration for each compartment	kg/m^3	$c[1\ldots7]$
mass for each compartment	kg	$m[1\ldots7]$
mobility for each compartment	kg/s	$Mob[1\ldots7]$
time when 95% of steady-state is reached	s	$tsteady$
average system decomposition rate	1/a	$System\,Decomposition$
half-life of the substance	a	$HalfLife$

The input parameters are described at the end of this section.

The model consists of the following equations:

$i = 0$
$Kcorr = 1$
IF compartment Water selected THEN
$\quad j = j + 1$
$\quad K[j] = 1$
$\quad V[j] = Water.Area \cdot Water.Depth$
$\quad Q_{in}[j] = Water.Q$
$\quad \lambda[j] = Deg_{Water}$
$\quad C_i[j] = Water.C_{in}$
END

IF compartment Air selected THEN
$\quad j = j + 1$
$\quad$IF $j > 1$ THEN
$\quad\quad K[j] = K_{AW}$
$\quad$ELSE
$\quad\quad K[j] = 1$

$$Kcorr = K_{AW}$$
END
$$V[j] = 0$$
IF compartment Water selected THEN
$$\quad V[j] = Water.Area$$
END
IF compartment Soil selected THEN
$$\quad V[j] = V[j] + Soil.Area$$
END
$$V[j] = V[j] \cdot Air.Height$$
$$Q_{in}[j] = Air.Q$$
$$\lambda[j] = Deg_{Air}$$
$$C_i[j] = Air.C_{in}$$
END

IF compartment Soil selected THEN
$$\quad j = j + 1$$
$$\quad K[j] = Soil.K_{BW}/Kcorr$$
$$\quad V[j] = Soil.Area \cdot Soil.Depth$$
$$\quad Q_{in}[j] = 0$$
$$\quad \lambda[j] = Deg_{Soil}$$
$$\quad C_i[j] = 0$$
END

IF compartment Sediment selected THEN
$$\quad j = j + 1$$
$$\quad K[j] = Sediment.K_{SW}/Kcorr$$
$$\quad V[j] = Sediment.Area \cdot Sediment.Depth$$
$$\quad Q_{in}[j] = 0$$
$$\quad \lambda[j] = Sediment.degra$$
$$\quad C_i[j] = 0$$
END

IF compartment Suspended matter selected THEN
$$\quad j = j + 1$$
$$\quad K[j] = Suspend.K_d \cdot (Suspend.rho/1000)/Kcorr$$
$$\quad V[j] = \frac{(Water.Depth \cdot Water.Area \cdot Suspend.Content)}{Suspend.rho}$$
$$\quad Q_{in}[j] = Water.Q \cdot Suspend.Content$$
$$\quad \lambda[j] = Deg_{Water}$$
$$\quad C_i[j] = Suspend.C_{in}$$
END

IF compartment Fish selected THEN
    $j = j + 1$
    $K[j] = BCF \cdot (Fish.rho/1000)/Kcorr$
    $V[j] = \dfrac{Water.Depth \cdot Water.Area \cdot Fish.Content}{Fish.rho}$
    $Q_{in}[j] = 0$
    $\lambda[j] = Fish.degra$
    $C_i[j] = 0$
END

IF compartment Plant selected THEN
    $j = j + 1$
    $K[j] = Plant.K_{PW}/Kcorr$
    $V[j] = \dfrac{Soil.Area \cdot Plant.Vegetation}{Plant.rho}$
    $Q_{in}[j] = 0$
    $\lambda[j] = Deg_{Plant}$
    $C_i[j] = 0$
END

$max = j$

$Input = Gen.Input$;beginning of the Level 2 equations
$$help1 = V[1] + \sum_{j=2}^{max} K[j] \cdot V[j]$$

$$c_{01} = \frac{Gen.InitialMass}{help1} \quad ;\text{compute initial value } t = 0$$

$$help2 = \sum_{j=1}^{max} \left(\frac{\lambda[j] \cdot V[j] \cdot K[j]}{86400} + Q_{in}[j] \cdot K[j] \right)$$

$a = help2/help1$

$help2 = Input/31.536 \cdot 10^6$;annual input
$$help2 = help2 + \sum_{j=1}^{max} (Q_{in}[j] \cdot Ci[j])$$

$b = help2/help1$

$ExistsSteadyState = TRUE$
IF $(a \leq 0)$ AND $(b \leq 0)$ THEN
    $tsteady = 0$;no input and no degradation
ELSE
    IF $(a \leq 0)$ OR $(b \leq 0)$ THEN
        $tsteady = \infty$;no input or no degradation
        $ExistsSteadyState = FALSE$
    ELSE

$tsteady = \ln(20)/a$;time when 95% of the steady-state is reached
   END
END

IF NOT *Gen.Time* = *unknown* THEN
   $t = 0$
   WHILE $t \leq$ *Gen.Time* $\cdot 31.536 \cdot 10^6$ DO
      IF $a \leq 0$ THEN
         IF $b \leq 0$ THEN
            $c[1] = c_{01}$
         ELSE
            $c[1] = c_{01} + b \cdot t$
         END
      ELSE

$$c[1] = c_{01} \cdot e^{-a \cdot t} + \frac{b}{a} \cdot \left(1 - e^{-a \cdot t}\right) \quad \text{;analytical solution of } y' = b + ya$$

      END
      $t = t +$ *Gen.Step* $\cdot 31.536 \cdot 10^6$
      $m[1] = V[1] \cdot c[1]$;mass in compartment 1
      $Mob[1] = c[1] \cdot Q_{in}[1]$
      FOR $j = 2$ TO max DO
         $c[j] = K[j] \cdot c[1]$;concentration at t
         $m[j] = V[j] \cdot c[j]$
         $Mob[j] = c[j] \cdot Q_{in}[j]$
      END
   END
END

IF *ExistsSteadyState* THEN   ;if steady-state exists
   IF *tsteady* > 0 THEN
      $c[1] = b/a$
   ELSE
      $c[1] = c_{01}$;no degradation and no input
   END
   $m[1] = V[1] \cdot c[1]$
   $Mob[1] = c[1] \cdot Q_{in}[1]$
   FOR $j = 2$ TO *max* DO
      $c[j] = K[j] \cdot c[1]$;concentration in compartment j
      $m[j] = V[j] \cdot c[j]$
      $Mob[j] = c[j] \cdot Q_{in}[j]$
   END

$$MassSum = \sum_{j=1}^{max} m[j]$$

IF $Input = 0$ THEN
   $SystemDecomposition = 0$
ELSE
   IF $MassSum = 0$ THEN
      $SystemDecomposition = 0$
   END
ELSE
   $SystemDecomposition = Input/MassSum$;compute system decomposition rate

END
IF $SystemDecomposition = 0$ THEN
   $HalfLife = 0$
ELSE
   $HalfLife = (\ln(2)/(Input = MassSum))$;compute half-life
END

with

Notation	Description	Unit
$K[1 \ldots 7]$	Distribution coefficient ref. Water or Air	–
$V[1 \ldots 7]$	Volume of each compartment	m^3
$\lambda[1 \ldots 7]$	Degradation rate of each compartment	$1/d$
$Q_{in}[1 \ldots 7]$	Thru-flux of each compartment	m^3/s
$C_i[1 \ldots 7]$	Concentration in thru-flux	kg/m^3
c_{01}	Initial concentration	kg/m^3
$Input$	Annual input	kg/a
$MassSum$	Sum of all masses in the compartments	kg
$Kcorr$	correction coefficient	–
max	Number of selected compartments	–
$a; b$	Aggregated input and degradation	–
$ExistsSteadyState$	$= TRUE$ if steady-state exists	–
t	Time step of calculation	s
$j, help1, help2$	Dummies	–

BCF, K_{AW}, Deg_{Water}, Deg_{Air}, Deg_{Plant} and Deg_{Soil} are $\rightarrow$ Substance Data [Page 221].

Error messages:

Condition	Reason
$Gen.Step > Gen.Time$	No reasonable computing is possible.

The following combinations are compulsory (combine a row with a column):

	Water	Air	Soil	Sediment	Suspended mat.	Fish	Plant
Water							
Air	○		○				
Soil	○	○					
Sediment	*						
Suspended matter	*						
Fish	*						
Plant		*	*				

○ = have to be combined with at least one of the marked compartments
* = have to be combined with all of the marked compartments

Description of input parameter:

Notation	Unit	Default	Warning range	Error range
Gen.InitialMass	kg	*unknown*	$[0; 1 \cdot 10^9]$	$[0; \infty[$
Precontamitation of the system.				
Gen.Input	kg/a	*unknown*	$[0; 1 \cdot 10^9]$	$[0; \infty[$
Annual input in the system.				
Gen.Time	a	1	$[1/365; 1 \cdot 10^9]$	$[1/365; 1 \cdot 10^9]$
Time period of the simulation; if unknown only the steady state is calculated.				
Gen.Step	a	0.05	$[1/365; 1 \cdot 10^4]$	$[1/365; 70]$
Stepsize for the result display.				
Water.Area	m²	$3.56963 \cdot 10^9$	$[0; 1 \cdot 10^{10}]$	$[0; 3; 5 \cdot 10^{14}]$
Sum of all water surfaces of the system.				
Water.Depth	m	2	$[0; 10000]$	$[0; 10000]$
Average depth of all waters in the system.				
Water.pH	–	6	$[2; 12]$	$[0; 14]$
pH of water; only used for standard pH-correction.				
Water.Q	m³/s	$1 \cdot 10^4$	$[0; 30000]$	$[0; \infty[$
Thruflux of water.				
Water.Cin	kg/m³	0	$[0; 500]$	$[0; 1300]$
Concentration of pollutant in water thruflux.				
Air.Height	m	6000	$[0; 20 \cdot 10^3]$	$[0; 20 \cdot 10^3]$
Height of compartment air.				

Air.Q	m^3/s	$6 \cdot 10^9$	$[0; \infty[$	$[0; \infty[$

Thruflux of polluted air.

Air.Cin	kg/m^3	0	$[0; 0.5]$	$[0; 2]$

Concentration of the pollutant in thruflux air.

Soil.Area	m^2	$3.56963 \cdot 10^{11}$	$[0; 510 \cdot 10^{12}]$	$[0; 1.6 \cdot 10^{14}]$

Soil surface of the system.

Soil.Depth	m	0.2	$[0; 3]$	$[0; 6300 \cdot 10^3]$

Average depth of the soil.

Soil.Density	kg/m^3	1500	$[0; 5000]$	$[0; 20000]$

Density of soil. Only used for the estimation of K_{BW}.

Soil.K$_d$	cm^3/g	*unknown*	$[0; 1 \cdot 10^6]$	$[0; \infty[$

Distribution coefficient of soilmatrix/water; only used for the estimation of K_{BW}. Estimation: See $\rightarrow$ K_d-Estimation [Page 266].

Soil.K$_{BW}$	–	*unknown*	$[0; 1 \cdot 10^6]$	$[0; \infty[$

See $\rightarrow$ K_{BW}-Estimation [Page 274]. Distribution coefficient soil/water.

Soil.theta	m^3/m^3	0.3	$[0; 1]$	$[0; 1]$

Water content in the pores of the soil; only used for the estimation of K_{BW}.

Soil.Epsilon	m^3/m^3	0.5	$[0; 1]$	$[0; 1]$

Fraction of total pore volume of dry soil mass; only used for the estimation of K_{BW}.

Soil.OC	kg/kg	0.02	$[0; 1]$	$[0; 1]$

Content of organic carbon in soil.

Soil.pH	–	6	$[2; 12]$	$[0; 14]$

pH of the soil; only used for the estimation of K_{BW}.

Sediment.Area	m^2	$2.498741 \cdot 10^9$	$[0; 510 \cdot 10^{12}]$	$[0; 510 \cdot 10^{12}]$

Area of the sediment.

Sediment.Depth	m	0.05	$[0; 1]$	$[0; 6300 \cdot 10^3]$

Average depth of the sediment.

Sediment.Density	kg/m^3	1000	$[0; 5000]$	$[0; 20000]$

Dry density of the sediment; only used for the estimation of K_{SW}.

Sediment.Degra	$1/d$	*unknown*	$[0; 36]$	$[0; \infty[$

Rate of degradation in the sediment.

Sediment.K$_d$	cm^3/g	*unknown*	$[0; 1 \cdot 10^6]$	$[0; \infty[$

Distribution coefficient of sedimentmatrix/water. Estimation: See $\rightarrow$ K_d-Estimation [Page 266].

Sediment.K$_{SW}$	–	*unknown*	$[0; 1 \cdot 10^6]$	$[0; \infty[$

See $\rightarrow$ K_{SW}-Estimation [Page 273]. Distribution coefficient total sediment/water.

Sediment.Theta m^3/m^3 0.5 [0; 1] [0; 1]
Water content of the sediment; only used for the standard correction of K_d.

Sediment.OC kg/kg 0.05 [0; 1] [0; 1]
Content of organic carbon in the soil; only used for the estimation of K_d.

Sediment.pH – 6 [2; 12] [0; 14]
pH of the sediment; only used for pH correction.

Suspend.Content kg/m^3 0.05 [0; 3] [0; 25000]
Average content of suspended matter in water.

Suspend.Cin kg/m^3 0 [0; 500] [0; 25000]
Concentration of the pollutant in the thruflux.

Suspend.Rho kg/m^3 1500 [0; 5000] [0; 25000]
Dry density of the suspended matter.

Suspend.K$_d$ cm^3/g *unknown* [0; $1 \cdot 10^6$] [0; ∞[
Distribution coefficient of suspended matter/water. Estimation: See → K_d-Estimation [Page 266].

Suspend.OC kg/kg 0.1 [0; 1] [0; 1]
Content of organic carbon; only used for the estimation of K_d.

fish.Content kg/m^3 0.5 [0; 10] [0; 3569.63]
Content of fish in the water.

fish.Rho kg/m^3 1000 [0; 1300] [0; 5000]
Average density of the fish.

fish.Degra 1/d *unknown* [0; 36] [0; ∞[
Average degradation in fish.

Plant.Vegetation kg/m^2 1 [0; 50] [0; 500]
Average vegetation density of the soil.

Plant.K$_{PW}$ – *unknown* [0; $1 \cdot 10^6$] [0; ∞[
See → K_{PW}-Estimation [Page 273]. Distribution coefficient total plant/water.

Plant.Rho kg/m^3 1000 [0; 2500] [0; 25000]
Fresh density of the plant.

Plant.Wp kg/kg 0.8 [0; 1] [0; 1]
Water content of the plant; only used for the estimation of K_{PW}.

Plant.Lp kg/kg 0.02 [0; 1] [0; 1]
Lipid content of the plant; only used for the estimation of K_{PW}.

9.6
Plant

Plant describes the uptake of chemical substances from soil and air into plants
→(Trapp and McFarlane 1995) [Page 172]. For concentration in roots equilibrium distribution with soil is assumed. Concentration in aerial plant parts
(mainly leaves) can be calculated with or without particle deposition. If the
plants are in the exponential growth phase, the dilution of the substance can
be found using the rate of growth. The equation of mass balance is solved
analytically.

Development over time of the total concentration, a table with the development
of the partial concentrations and following results are shown:

Description	Unit	Notation
Steady-state concentration of the root	kg/m^3	C_r
Total conductance	m/s	G
Total loss rate by volatilization	1/d	$a1_d$
Loss by volatilization	%	a_{vol}
Total degradation rate plant	1/d	$a2_d$
Loss by degradation	%	a_{deg}
Total loss rate	1/d	a_d
Absorption from soil	kg/(m^{3}d)	$b1_d$
Uptake from soil	%	b_{asoil}
Absorption from air	kg/(m^{3}d)	$b2_d$
Uptake from air	%	b_{aair}
Total absorption	kg/(m^{3}d)	b_d
Steady-state concentration	kg/m^3	$ssConc$
Time to steady-state (95%)	d	$tss95$

Plant carries out the following calculations:

$$C_r = K_{RW} \cdot C_{SW}$$

If selected, the total conductance will be calculated in the following way:

$$Gk = \frac{10^{0.704 \cdot \log K_{OW} - 11.2}}{K_{AW}}$$

$$Ga = 0.005 \cdot \sqrt{\frac{300}{Mol}}$$

$$Gc = \frac{1}{\frac{1}{Gk} + \frac{1}{Ga}}$$

$$Ew = 610.7 \cdot 10^{\frac{7.5 \cdot (temp - 273.15)}{237 + (temp - 273.15)}}$$

$$d_{ws} \;=\; \frac{Ew}{461 \cdot temp}$$

$$Gw \;=\; \frac{1000 \cdot Transpirationstream}{(d_{ws} - d_{ws} \cdot humid) \cdot 86400}$$

$$Gs \;=\; Gw \cdot \sqrt{\frac{18}{Mol}} \cdot \frac{1}{Area_L}$$

$$G \;=\; Gc + Gs$$

At this position G is known as calculated or input conductance. It depends on the switch "Total conductance?" in the dialog box "model options".

$$a1 \;=\; \frac{Area_L \cdot G}{K_{LA} \cdot Volume_L}$$

$$a2 \;=\; \frac{Deg_{Hant}}{86400}$$

$$a3 \;=\; \frac{growth_r}{86400}$$

$$a \;=\; a1 + a2 + a3$$

$$b1 \;=\; \frac{Transpirationstream \cdot TSCF \cdot C_{solu}}{Volume_L \cdot 86400}$$

$$V_{gw} \;=\; \frac{rain}{K_{AW} \cdot 8.64 \cdot 10^7}$$

$$b2 \;=\; \frac{(G + 0.5 \cdot V_{gw}) \cdot (1 - f_p) \cdot Area_L \cdot C_{air}}{Volume_L}$$

$$b \;=\; b1 + b2$$

The output consists of the results from a1, a2, b1, b2 and b in $1/d$. They are multiplied by the factor 86 400. With this one gets the values $a1_d$, $a2_d$, $b1_d$, $b2_d$ and b_d.

$$a_{vol} \;=\; \frac{a1 \cdot 100}{a1 + a2}$$

$$a_{deg} \;=\; \frac{a2 \cdot 100}{a1 + a2}$$

$$b_{asoil} \;=\; \frac{b1 \cdot 100}{b}$$

$$b_{aair} \;=\; \frac{b2 \cdot 100}{b}$$

$$a_d \;=\; (a1 + a2) \cdot 86400$$

$$ssConc \;=\; \frac{b}{a}$$

$$tss95 \;=\; \frac{\ln(20)}{a} \cdot \frac{1}{86400}$$

Calculation under consideration of particles:

$$v_{pw} = \frac{washoutRatio \cdot rain}{8.64 \cdot 10^7}$$

$$bp = (0.5 \cdot V_{pd} + 0.5 \cdot V_{pw}) \cdot f_p \cdot Area_L \cdot \frac{C_{air}}{Volume_L}$$

$$ap = \frac{weathering_r + growth_r}{86400}$$

The following calculates the concentrations for time $Calc_t$.

$$C_{L,t} = C_{L,0} \cdot e^{-a \cdot Calc_t \cdot 86400} + \frac{b}{a} \cdot (1 - e^{-a \cdot Calc_t \cdot 86400})$$

In the case of calculation with consideration of particles:

$$C_{P,t} = C_{P,0} \cdot e^{-ap \cdot Calc_t \cdot 86400} + \frac{bp}{ap} \cdot (1 - e^{-ap \cdot Calc_t \cdot 86400})$$

$$C_t = C_{L,t} + C_{P,t}$$

with

Notation	Description	Unit
$Calc_t$	Moment of calculation	d
C_t	Total concentration at moment $Calc_t$	kg/m^3
$C_{P,t}$	Concentration at particle at moment $Calc_t$	kg/m^3
$C_{L,t}$	Concentration in leaves at moment $Calc_t$	kg/m^3
$a, a1, a2, a3$	Internal rates	1/s
$b, b1, b2$	Internal rates	1/s
bp, ap	Interal rates for particles	1/s
Gc	Conductance by uptake from cuticle	m/s
Gs	Conductance by uptake from stomata	m/s
Gk	Conductance of cuticle	m/s
Ga	Conductance of atmospheric boundary layer	m/s
Gw	Conductance for water in whole plant	m/s
d_{ws}	Vapor density	kg/m^3
Ew	Saturation vapor pressure of water	Pa
V_{gw}	Gas deposition velocity, wet	m/s
V_{pw}	Particle deposition velocity, wet	m/s
$washoutRatio$	Efficiency of washout of precipitation (here constant $= 2 \cdot 10^5$)	(kg/m^3 rain)/ (kg/m^3 air)

Values $\log K_{OW}$, K_{AW}, Mol and Deg_{Hant} belong to →Substance Data [Page 221].

Hints:

Condition	Reason
After calculation $b = 0$	There is no uptake.

Error messages:

Condition	Reason
$Volume_L$ is 0.	The model is not applicable.
During the calculation of total conductance $K_{AW} = 0$.	Calculation of total conductance for $K_{AW} = 0$ is not possible.
After calculation $a = 0$.	Due to the missing loss rate the model is not applicable.
After calculation $ap = 0$	Growth rate of plant and weathering rate of particles cancel each other. The calculation with particle consideration is not possible.

The type of calculation can be chosen with the switch "Calculation". Depending on its position, the concentration of particles will either be calculated or not considered.

The switch "Total Conductance" determines the calculation of the total conductance. Depending on its position, the total conductance will either be calculated within the model or the input conductance will be used.

Description of input parameter:

Notation	Unit	Default	Warning range	Error range
start$_t$ Starting time of the simulation.	d	0	[0; 180[	[0; ∞[
duration$_t$ Duration of simulation.	d	60	]0; 180[	]0; ∞[
step$_t$ Time steps of simulation.	d	1	]0; 180[	]0; ∞[
C$_{L,0}$ Start concentration in leaves.	kg/m^3	0	[0; 500]	[0; 25000]
Area$_L$ Total leaf surface area.	m^2	4	]0; $1 \cdot 10^6$[	]0; ∞[
Volume$_L$ Volume of the aerial plant parts.	m^3	0.002	]0; 5000[	]0; ∞[

growth_r   1/d   0.035 $[0; 1]$ $]-\infty; \infty[$
Growth rate of plant in exponential phase.

Transpiration m^3/d 0.001 $[0; 2500[$ $[0; \infty[$
stream
Transpiration stream.

C_{Air} kg/m^3 *unknown* $[0; 0.5]$ $[0; 2]$
Total concentration in air.

rain   mm/d   3 $[0; 2000]$ $[0; \infty[$
Rate of precipitation.

TSCF   –   *unknown* $[0; 1[$ $[0; \infty[$
Concentration factor for transpiration stream. Estimation: See →Estimation of TSCF [Page 270].

f_p –   *unknown* $[0; 1]$ $[0; 1]$
Particulate fraction. Estimation: See →Estimation of Particulate fraction [Page 270].

$C_{P,0}$ kg/m^3 0 $[0; 500]$ $[0; 25000]$
Start concentration of particles on leaves.

V_{pd} m/s $5 \cdot 10^{-4}$ $[1 \cdot 10^{-5}; 1]$ $[0; 3 \cdot 10^8]$
Velocity of particle deposition.

weathering_r   1/d   0.05 $]0; 36]$ $]0; \infty[$
Weathering rate of particles.

G   m/s   unknown $]0; 1 \cdot 10^{-2}]$ $]0; \infty[$
Total conductance leaf-air.

C_{SW} kg/m^3 *unknown* $[0; 500]$ $[0; 25000]$
Concentration in the soilwater. Estimation: See →Estimation of concentration in the soilwater [Page 271].

K_{RW} m^3/m^3 *unknown* $[0; 1 \cdot 10^6]$ $[0; \infty[$
Distribution coefficient root/water. Estimation: See → K_{RW} Estimations [Page 271].

l_{roots} kg/kg   0.01 $]0; 1]$ $[0; 1]$
Lipid content of roots (fresh weight).

Water_roots   kg/kg   0.8 $]0; 1]$ $[0; 1]$
Water content of the roots (fresh weight).

D_{roots} kg/m^3 1000 $[100; 2000]$ $[0; 25000]$
Density of roots (fresh weight).

K_{LA} –   unknown $]0; 1 \cdot 10^{12}[$ $]0; \infty[$
Distribution coefficient leaves/atmosphere. Estimation: See → K_{LA} Estimations [Page 272].

| D_{plant} | kg/m^3 | 500 | [100; 2000] | [0; 25000] |

Density of aerial plant parts (fresh weight).

| l_{plant} | kg/kg | 0.03 |]0; 1] | [0; 1] |

Lipid content of aerial plant parts (fresh weight).

| Water$_{plant}$ | kg/kg | 0.8 |]0; 1] | [0; 1] |

Water content of aerial plant parts (fresh weight).

| humid | – | 0.5 | [0; 0.9] | [0; 1[|

Relative humidity.

| temp | K | 293.15 | [273.15; 293.15] |]0; 10000[|

Temperature.

| Porosity | – | 0.5 |]0; 1] |]0; 1] |

Porosity of soil.

| D_{soil} | kg/m^3 | 1300 | [500; 3000] | [0; 25000] |

Density of soil.

| pore | – | 0.3 |]0; 1] |]0; 1] |

Fraction of water-filled pores.

| C_{soil} | kg/m^3 | *unknown* | [0; 500] | [0; 25000] |

Concentration in soil.

| K_d | cm^3/g | *unknown* | [0; $1 \cdot 10^6$] | [0; ∞[|

Distribution coefficient matrix/water. Estimation: See → K_d-Estimation [Page 266].

| OC | kg/kg | 0.04 | [0; 1] | [0; 1] |

Organic carbon content of soil. It is used for the estimation of K_d.

| pH | – | 6 | [2; 12] | [0; 14] |

pH-value of soil-water for the estimation of K_d. In addition, it is also used for the standard correction of K_{AW}.

9.7
Plume

This model describes the atmospheric transport of chemical substances after point source emission. Atmospheric and meteorological conditions are taken as constant. Plume includes dilution by atmospheric dispersion (Gaussian type). The model is based on the three-dimensional dispersion-advection equation.

In addition, desposition from steady-state concentrations and rates of deposition are calculated, as well as potential dose rates from average inhalation rates.

With Plume the expected concentration can be calculated at a special point or in the main direction of the wind. In the latter case, tables for distance/deposition and distance/concentration are shown. For the calculation at one point only the following results will be shown:

Description	Unit	Notation
Concentration at the point	kg/m^3	$C_{x,y,z}$
Total deposition at the point	kg/(m$^2 \cdot a$)	$dep_{x,y,0}$
Annual inhalation dose at the point	kg/a	$inhal_{dose}$
Distance to point of max. concentration	m	$point_{max}$
Maximal concentration	kg/m^3	C_{max}
Maximal deposition	kg/m$^2 \cdot a$	dep_{max}
Maximal annual inhalation dose	kg/a	$InhalDose_{max}$

This model carries out the following computations:

$$v_{dep,dry} = (1 - f_p) \cdot VdGasDry + f_p \cdot VdParDry$$

$$v_{dep,wet} = \left(\frac{1 - f_p}{K_{AW}} + f_p \cdot washoutRatio \right) \cdot \frac{rain}{8.64 \cdot 10^7}$$

$$v_{dep} = v_{dep,dry} + v_{dep,wet}$$

Computation of concentration at one point:

$$\sigma_y = a \cdot x^p$$

$$\sigma_z = b \cdot x^q$$

$$C_{x,y,z} = \frac{input}{(2 \cdot \pi \cdot \sigma_y \cdot \sigma_z \cdot v_{Wind} \cdot 86400)}$$
$$\cdot e^{\frac{-y^2}{2 \cdot \sigma_y^2}} \cdot \left(e^{\frac{-(z - release_{height})^2}{2 \cdot \sigma_z^2}} + e^{\frac{-(z + release_{height})^2}{2 \cdot \sigma_z^2}} \right)$$

With points on the ground ($z = 0$):

$$dep_{x,y,0} = v_{dep} \cdot C_{x,y,z} \cdot 86400 \cdot 365$$

$$inhalDose = inhal \cdot 365 \cdot C_{x,y,z}$$

For the calculation of the main direction of the wind:

$$point_{max} = \left(\frac{(q + p) \cdot b^2}{release_{height}^2 \cdot q} \right)^{-0.5/q}$$

$$\sigma_y = a \cdot point_{max_p}$$

$$\sigma_z = b \cdot point_{max_q}$$

$$C_{max} = \frac{input}{(\pi \cdot \sigma_y \cdot \sigma_z \cdot v_{Wind} \cdot 86400)} \cdot e^{\frac{-release_{height}^2}{2 \cdot \sigma_z^2}}$$

$$dep_{max} = v_{dep} \cdot 86400 \cdot 365 \cdot C_{max}$$

$$inhalDose_{max} = inhal \cdot 365 \cdot C_{max}$$

For the points in the main direction of the wind specified by start, end and step size:

$$\sigma_y = a \cdot x^p$$
$$\sigma_z = b \cdot x^q$$

$$C_x = \frac{input}{(\pi \cdot \sigma_y \cdot \sigma_z \cdot v_{Wind} \cdot 86400)} \cdot e^{\frac{-release_{height}^2}{2\sigma_z^2}}$$
$$dep_x = v_{dep} \cdot 86400 \cdot 365 \cdot C_x$$

with

Notation	Description	Unit
$v_{dep,dry}$	Total dry deposition velocity	m/s
$v_{dep,wet}$	Total wet deposition velocity	m/s
v_{dep}	Total deposition velocity	m/s
washoutRatio	Washout coefficient of precipitation (here constant $= 2 \cdot 10^5$)	(kg/m^3 precipitation)/ (kg/m^3 air)
σ_y	Total lateral dispersion coefficient	–
σ_z	Total vertical dispersion coefficient	–

K_{AW} belongs to → Substance Data [Page 221].

Error messages:

Condition	Reason
K_{AW} of substance is 0.	The model is not applicable for a substance with $K_{AW} = 0$.
With $start_x > end_x$ or $step_x > (start_x - end_x)$ (checked in input dialog)	No sensible calculation is possible.

The type of calculation can be chosen using the switch "Calculation". It can either relate to the calculation of concentrations at one point or in the main direction of the wind.

Description of input parameter:

Notation	Unit	Default	Warning range	Error range
input	kg/d	1	$[0; \infty[$	$[0; \infty[$
Release rate of substance.				
inhal	m^3/d	20	$[2; 40]$	$[0; \infty[$
Average inhalation rate.				
release$_{height}$	m	50	$[0; 50]$	$[0; 2200]$
Effective height of release.				
f_p	–	*unknown*	$[0; 1]$	$[0; 1]$
Adsorbed fraction. Estimation: See $\rightarrow$ Estimation of adsorbed fraction [Page 270].				
v_{Wind}	m/s	5	$[0; 20]$	$[0; 3 \cdot 10^8]$
Wind speed in effective height of release.				
rain	mm/d	2.1	$[0; 2000]$	$[0; \infty[$
Average precipitation rate.				
VdParDry	m/s	0.01	$[0; 1]$	$[0; 3 \cdot 10^8]$
Deposition velocity of dry particle.				
VdGasDry	m/s	*unknown*	$[0; 1]$	$[0; 3 \cdot 10^8]$
Deposition velocity of dry gas. Estimation: See $\rightarrow$ VdGas-Estimation [Page 266].				
a	–	0.64	$[0.170; 1.503]$	$]0; 10]$
Lateral dispersion coefficient a; Default value for effective height of release $h \leq 50$ m and neutral stability class following the German "TA-Luft" III/1, Appendix C.				
p	–	0.784	$[0.710; 1.296]$	$]0; 10]$
Lateral dispersion exponent p; Default value for effective height of release $h \leq 50$ m and neutral stability class following the German "TA-Luft" III/1, Appendix C.				
b	–	0.215	$[0.051; 0.717]$	$]0; 10]$
Vertical dispersion coefficient b; Default value for effective height of release $h \leq 50$ m and neutral stability class following the German "TA-Luft" III/1, Appendix C.				
q	–	0.885	$[0.486; 1.317]$	$]0; 10]$
Vertical dispersion exponent q; Default value for effective height of release $h \leq 50$ m and neutral stability class following the German "TA-Luft" III/1, Appendix C.				
start$_x$	m	0	$[0; 50 \cdot 10^3[$	$[0; 40 \cdot 10^6[$
Starting point of vectorial calculation.				

end$_x$	m	10000	$]0; 50 \cdot 10^3]$	$]0; 40 \cdot 10^6]$
Ending point of vectorial calculation.				
step$_x$	m	100	$]0; 10 \cdot 10^3]$	$]0; 10 \cdot 10^6]$
Step size of vectorial calculation.				
x	m	1000	$]0; 50 \cdot 10^3]$	$]0; 40 \cdot 10^6]$
X-Coordinate of one-point calculation.				
y	m	10	$[-50 \cdot 10^3; 50 \cdot 10^3]$	$[-40 \cdot 10^6; 40 \cdot 10^6]$
Y-Coordinate of one-point calculation.				
z	m	0	$[-1100; 2000]$	$[-20 \cdot 10^3; 20 \cdot 10^3]$
Z-Coordinate of one-point calculation.				

9.8
Soil

Soil describes the transport and whereabouts of pollutants in soil. It includes three analytical solutions for simulating a pulse input, a contaminated layer and a continuous injection. Furthermore, the steady-state solution is calculated for continuous injection.

Transport velocity and the total diffusion/dispersion coefficient can be estimated or entered manually. In this way this model can be used for transport processes beyond soil.

Output consists of the calculated concentration, fractions of the substance in aqueous phase, gaseous phase and in the soil matrix. Furthermore, the vertical transport velocity of the substance and total diffusion/dispersion coefficient under the evaluation of the fractions of substance in aqueous and gaseous phase can also be calculated.

Decription	Unit	Notation
Concentration	kg/m^3	C_1
Concentration (steady-state)	kg/m^3	C_2
Fraction in aqueous phase	–	f_W
Fraction in gaseous phase	–	f_G
Fraction in soil matrix	–	f_M
Transport velocity	m/d	u
Dispersion coefficient	m^2/d	D

With this model the $\rightarrow K_{AW}$ standard correction is carried out automatically beforehand [Page 275] . The following computations are carried out:

$$OM = 1.724 \cdot OC$$

$$Density = (1 - Porosity) \cdot (OM \cdot 1400 + (1 - OM) \cdot 2650)$$

$$K_{MW} = K_d \cdot Density \cdot 10^{-3}$$

$$f_W = \frac{pore}{K_{MW} + pore + (Porosity - pore) \cdot K_{AW}}$$

$$f_G = \frac{(Porosity - pore) \cdot K_{AW}}{K_{MW} + pore + (Porosity - pore) \cdot K_{AW}}$$

$$f_M = \frac{K_{MW}}{K_{MW} + pore + (Porosity - pore) \cdot K_{AW}}$$

Soil 1 (pulse input):

$$C_1 = \frac{m}{\sqrt{4 \cdot \pi \cdot D \cdot t}} \cdot e^{\frac{-(z - u \cdot t)^2}{4 \cdot D \cdot t}} \cdot e^{-Deg_{Soil} \cdot t}$$

Soil 2 (contaminated layer):

$$h = h/2$$

$$C_1 = 0.5 \cdot C_0 \cdot \left(erf \frac{z - ut + h}{\sqrt{4 \cdot D \cdot t}} - erf \frac{z - ut - h}{\sqrt{4 \cdot D \cdot t}} \right) \cdot e^{-Deg_{Soil} \cdot t}$$

Soil 3 (continuous injection):

$$\beta = \sqrt{\frac{u^2}{4 \cdot D^2} + \frac{Deg_{Soil}}{D}}$$

$$C_1 = \frac{C_0}{2} \cdot e^{\frac{u \cdot z}{2 \cdot D}} \cdot \left(e^{-z \cdot \beta} \cdot erfc \frac{z - t \cdot \sqrt{u^2 + 4 \cdot Deg_{Soil} \cdot D}}{\sqrt{4 \cdot D \cdot t}} \right.$$

$$\left. + e^{z \cdot \beta} \cdot erfc \frac{z + t \cdot \sqrt{u^2 + 4 \cdot Deg_{Soil} \cdot D}}{\sqrt{4 \cdot D \cdot t}} \right)$$

$$C_2 = C_0 \cdot e^{\frac{u \cdot z}{2 \cdot D} - z \cdot \beta}$$

with

Notation	Description	Unit
K_{MW}	Distribution coefficient	–
$D_{W,eff}$	Effective diffusion coefficient water	m^2/d
$D_{G,eff}$	Effective diffusion coefficient gas	m^2/d
$D_{W,a}$	Apparent dispersion/diffusion coeff.	m^2/d
OM	Content of organic matter	–
$Density$	Density of soil	kg/m^3

The remaining notations are → Substance Data [Page 221].

Description of input parameter:

Notation	Unit	Default	Warning range	Error range
Precipitation	mm/d	2.1	[0; 2000]	[0; ∞[
Amount of precipitation per day.				
Evaporation	mm/d	1.6	[0; 1999]	[0; ∞[
Evaporation per day.				
Runoff	mm/d	0.2	[0; 1999]	[0; ∞[
Runoff from soil surface per day.				
u	m/d	*unknown*	[0; 1]	[0; 3 · 10^8]
Vertical transport velocity of substance. Estimation: See →Estimation of transport velocity [Page 269].				
pore	m^3/m^3	0.3	]0; 1]	]0; 1]
Fraction of water-filled pores.				
Porosity	m^3/m^3	0.5	]0; 1]	]0; 1]
Porosity of soil.				
L$_{disp}$	m	0.05	[0; 100]	[0; ∞[
Length of dispersion.				
D	m^2/d	*unknown*	]0; 100]	]0; ∞[
Total dispersion/diffusion coefficient under evaluation of fractions of substance in aqueous and gaseous phase. Estimation: See →Estimation of total dispersion coefficient [Page 269].				
K$_d$	cm^3/g	*unknown*	[0; 1 · 10^6]	[0; ∞[
Distribution coefficient matrix/water. Estimation: See → K$_d$-Estimation [Page 266].				
OC	kg/kg	0.02	[0; 1]	[0; 1]
Organic carbon content of soil.				
pH	–	6	[2; 12]	[0; 14]
pH-value of aqueous phase.				
Inputfunction	–	*Soil1*	–	–
Type of model (Soil 1, Soil 2, Soil 3).				
m	kg/m^2	0.001	[0; 500]	[0; 1300]
Amount of input (only Soil 1).				
C$_0$	kg/m^3	0.1	[0; 500]	[0; 1300]
Concentration at beginning (only Soil 2 and 3).				
z	m	2	]0; 50]	]0; 6300 · 10^3]
Depth of layer under consideration.				
h	m	0.25	]0; 50]	]0; 6300 · 10^3]
Thickness of contaminated layer (only Soil 2).				
t	d	160	]0; 36500]	]0; ∞[
Period of time in days.				

9.9
Water

The model Water describes a continuous input by sewers into a water body. The water and mass flow are assumed to be stationary. The sorption to particles and sediment is computed vía partition coefficients. This model results in a one-dimensional description of the concentration in the water body (fluid).

The output is divided into three groups: General results, mass balance and a concentration profile:

Description	Unit	Notation
Total amount in water segment	kg	$mass_{total}$
Emission per day	kg/d	$mass_{input}$
Concentration fish at the start of the fluid	kg/m³	$C_{fish,max}$
Concentration fish at the end of the fluid	kg/m³	$C_{fish,min}$
Concentration sediment at the start of the fluid	kg/m³	$C_{sedi,max}$
Concentration sediment at the end of the fluid	kg/m³	$C_{sedi,min}$
Total elimination rate	1/d	$total_{elim}$
Sedimentation rate	1/d	sed_{rate}
Volatility rate	1/d	$volat$
Dilution factor	–	$dilution$
Concentration at the start of the fluid	kg/m³	C_0
Concentration at the end of the fluid	kg/m³	C_{end}
Mass flow by volatility	kg/d	$mass_{volat}$
Mass flow by sedimentation	kg/d	$mass_{sedim}$
Mass flow by degradation	kg/d	$mass_{degra}$
Mass flow by advection	kg/d	$mass_{advec}$

The model carries out the following calculations. With the model the $\rightarrow K_{AW}$ Standard correction is automatically carried out beforehand [Page 275].

$$cross_{section} = width \cdot depth$$

$$flow_{river} = vFlow \cdot cross_{section}$$

$$dilution = \frac{flow_{river}}{flowSew}$$

$$C_0 = \frac{conc_{sew}}{dilution}$$

$$frac_{partic} = \frac{suspM \cdot 10^{-3} \cdot K_d}{1 + suspM \cdot 10^{-3} \cdot K_d}$$

$$frac_{water} = \frac{1}{1 + suspM \cdot 10^{-3} \cdot K_d}$$

$$sed_{rate} = \frac{frac_{partic} \cdot 10^{-3} \cdot sedRat \cdot sedDen \cdot (1 - sedPor)}{365 \cdot depth \cdot suspM}$$

$$elim_{rate} = Deg_{Water} + volat + sed_{rate}$$

$$C_{end} = C_0 \cdot e^{\frac{-elim_{rate} \cdot length}{vFlow \cdot 86400}}$$

$$K_{SW} = \frac{K_d \cdot sedDen}{1000} + sedPor$$

$$mass_{total} = \frac{flow_{river} \cdot 86400 \cdot C_0 \cdot \left(1 - e^{\frac{-elim_{rate} \cdot length}{vFlow \cdot 86400}}\right)}{elim_{rate}}$$

$$mass_{input} = 86400 \cdot C_0 \cdot flow_{river}$$

$$C_{fish,max} = BCF \cdot C_0 \cdot frac_{water}$$

$$C_{fish,min} = BCF \cdot C_{end} \cdot frac_{water}$$

$$C_{sedi,max} = C_0 \cdot frac_{water} \cdot K_{SW}$$

$$C_{sedi,min} = C_{end} \cdot frac_{water} \cdot K_{SW}$$

$$mass_{volat} = volat \cdot mass_{total}$$

$$mass_{sedim} = sed_{rate} \cdot mass_{total}$$

$$mass_{degra} = Deg_{Water} \cdot mass_{total}$$

$$mass_{advec} = C_{end} \cdot flow_{river} \cdot 86400$$

with

Notation	Description	Unit
$cross_{section}$	Average cross-section of fluid	m^2
$flow_{river}$	Average flow of fluid	m^3/s
$frac_{water}$	Concentration water/Total concentration	–
$frac_{partic}$	Concentration sorped/Total concentration	–
K_{SW}	Concentration sediment/Concentration water	–

Protolysis, pH, pKa, K_{AW}, Deg_{Water} and BCF are $\rightarrow$ Substance Data [Page 221]. The first three are only required for the $\rightarrow$ K_{AW} Standard correction [Page 275].

Description of input parameter:

Notation	Unit	Default	Warning range	Error range
depth	m	3	$[0.1; 1000]$	$[1 \cdot 10^{-3}; 11000]$
Average depth of water body.				
width	m	330	$[2; 20 \cdot 10^3]$	$[1 \cdot 10^{-3}; 40 \cdot 10^6]$
Average width of water body.				
length	m	$200 \cdot 10^3$	$[1 \cdot 10^3; 6300 \cdot 10^3]$	$[10; 40 \cdot 10^6]$
Length of water body (maximum is the length of the Amazon = 6 300 km).				
OC	kg/kg	0.04	$[0; 1]$	$[0; 1]$
Organic carbon content of the sediment and the particles in the water.				
K_d	cm^3/g	*unknown*	$[0; 1 \cdot 10^6]$	$[0; \infty[$
Partition coefficient sediment matrix/sediment water. Estimation: See → K_d-Estimation [Page 266].				
pH	–	6	$[2; 12]$	$[0; 14]$
pH-value of water body.				
vFlow	m/s	1	$[1 \cdot 10^{-3}; 100]$	$[1 \cdot 10^{-3}; 3 \cdot 10^8]$
Flow velocity of the water body.				
vWind10	m/s	1	$[0; 20]$	$[0; 3 \cdot 10^8]$
Average wind velocity at a height of 10 m.				
flowSew	m^3/s	0.1	$[1 \cdot 10^{-5}; 126 \cdot 10^{12}]$	$]0; \infty[$
Average water input by sewers.				
conSew	kg/m^3	$10 \cdot 10^{-3}$	$]0; 500]$	$]0; 25000]$
Average concentration of substance in the sewers.				
sedPor	m^3/m^3	0.6	$[0; 1]$	$[0; 1]$
Porosity of the sediment.				
sedDen	kg/m^3	2000	$[0; 500]$	$[0; 25000]$
Density of dry sediment.				
sedRat	mm/a	10	$[0; 50]$	$[0; 11000 \cdot 10^3]$
Sedimentation of bound matter.				
suspM	kg/m^3	0.1	$]0; 3]$	$]0; 25000]$
Particle content in water body.				
segNum	–	50	$[10; 500]$	$[1; 500]$
Number of steps for graphical output.				
volat	1/d	*unknown*	$[0; 36]$	$[0; \infty[$
Volatility rate. Estimation: See →Volatility Estimations [Page 266].				

Chapter 10:
Estimation Functions

This chapter contains brief descriptions of all estimation functions available within this program. General information about parameter estimation is also included. Estimation functions can either be simple regressions or (especially with the estimation functions of model parameters) complex processes.

The application of estimations is useful if some of the needed parameters are not known. However, the use of estimated parameters as values in calculations results in the uncertainty of the reliability of the output. Estimated values should therefore not be used in a subsequent estimation. CemoS rejects mutual estimations of parameters, or more general, all circular estimations.

In general, the estimation functions prompt a warning or hint if a value is very inexact. The ranges and units are not mentioned here, since they were already dealt with in the model or substance descriptions. If a value has been estimated from other parameters, the calculation is automatically repeated if modifications to these parameters occur.

10.1
Estimation Functions for Substance Data

See → Estimate Input Data [Page 210] for a description of the practical use of estimation functions and automatic estimations.

10.1.1
Estimation of Molar Mass from Sum Formula

The molar mass $[g/mol]$ is estimated using the sum formula. Blanks before stochiometric coefficients in the sum formula are not permitted. Furthermore, it is possible to use brackets to group elements together. The stochiometric coefficient of a group of elements directly follows the closing bracket; it may not be above 10 000. You may, however, nest brackets.

The following values are used to estimate the molecular weight from the sum formula:

Ac 227	Cr 51.9961	K 39.0983	Pd 106.42	Tc 98
Ag 107.8682	Cs 132.9054	Kr 83.80	Pm 145	Te 127.60
Al 26.98154	Cu 63.546	La 138.9055	Po 209	Th 232.0381
Am 243	Dy 162.50	Li 6.941	Pr 140.9077	Ti 47.88
Ar 39.948	Er 167.26	Lr 260	Pt 195.08	Tl 204.383
As 74.9216	Es 252	Lu 174.967	Pu 244	Tm 168.9342
At 210	Eu 151.96	Md 258	Ra 226	U 238.0289
Au 196.9665	F 18.998403	Mg 24.305	Rb 85.4678	Unh 263
B 10.811	Fe 55.847	Mn 54.9380	Re 186.207	Unp 262
Ba 137.33	Fm 257	Mo 95.94	Rh 102.9055	Unq 261
Be 9.01218	Fr 223	N 14.0067	Rn 222	Uns 262
Bi 208.9804	Ga 69.723	Na 22.98977	Ru 101.07	V 50.9415
Bk 247	Gd 157.25	Nb 92.9064	S 32.066	W 183.85
Br 79.904	Ge 72.59	Nd 144.24	Sb 121.75	Xe 131.29
C 12.011	H 1.00794	Ne 20.179	Sc 44.95591	Y 88.9059
Ca 40.078	He 4.00260	Ni 58.69	Se 78.96	Yb 173.04
Cd 112.41	Hf 178.49	No 259	Si 28.0855	Zn 65.38
Ce 140.12	Hg 200.59	O 15.9994	Sm 150.36	Zr 91.224
Cf 251	Ho 164.9304	Os 190.2	Sn 118.710	
Cl 35.453	I 126.9045	P 30.97376	Sr 87.62	
Cm 247	In 114.82	Pa 231.0359	Ta 180.9479	
Co 58.9332	Ir 192.22	Pb 207.2	Tb 158.9254	

10.1.2
Estimation of Partition Coefficient Air/Water from Molar Mass, Water Solubility and Vapor Pressure

The patition coefficient air/water, K_{AW}, is estimated using water solubility, ws, and the vapor pressure, vp. This equation is applicable if the ratio between water solubility and molar mass, M, is smaller than 1 [mol/l].

If the vapor pressure is not known, a value of 1 013 hPa is assumed if the boiling point is below 20 degrees Celsius.

$$K_{AW} = \frac{M \cdot vp}{ws \cdot 1000 \cdot R \cdot 293.15}$$

with R: universal gas constant $= 8.314 \frac{J}{mol \cdot K}$ Molar mass, vapor pressure and water solubility are → Substance Data [Page 221].

With some models K_{AW} will be corrected by the → K_{AW} Standard Correction [Page 275].

10.1.3
Estimations of the Partition Coefficient Organic Carbon/Water

By Karickhoff (1981)

The partition coefficient of organic carbon to water, K_{OC}, is estimated using $\log K_{OW}$. The equation is only applicable for nondissociating organic chemicals; it was developed for five polycylic aromatic hydrocarbons and several sediments ($r^2 = 0.994$), which had a $\log K_{OW}$ of between 1.0 and 6.72.

$$K_{OC} = 0.411 \cdot 10^{\log K_{OW}}$$

The partition coefficient $\log K_{OW}$ is part of the $\rightarrow$ Substance Data [Page 221]. The reverse estimation function is $\rightarrow$ Estimation of partition coefficient octanol/water by Karickhoff [Page 263].

By Schwarzenbach and Westall (1981)

The partition coefficient of organic carbon to water, K_{OC}, is estimated using $\log K_{OW}$. The equation is only applicable for nondissociating organic chemicals; it was developed by examining a series of alcyliric and halogenic benzole and diverse soils.

$$K_{OC} = 10^{(0.72 \cdot \log K_{OW} + 0.49)}$$

The partition coefficient $\log K_{OW}$ is part of $\rightarrow$ Substance Data [Page 221]. The reverse estimation function is $\rightarrow$ Estimation of partition coefficient octanol/water by Schwarzenbach and Westall [Page 263].

From water solubility; by Kenaga and Goring (1980)

The partition coefficient of organic carbon to water, K_{OC}, is estimated using the water solubility, ws. A hint will be prompted if the water solubility does not fall within the valid range $[0.5 \cdot 10^{-6}; 1000]$ [kg/m^3]:

$$K_{OC} = 10^{(3.64 - 0.55 \cdot \log(ws \cdot 1000))}$$

The water solubility is part of the $\rightarrow$ Substance Data [Page 221].

From water solubility; by Chiou (1979)

The partition coefficient of organic carbon to water, K_{OC}, is estimated using the water solubility, ws.

$$K_{OC} = 10^{(4.273 - 0.686 \cdot \log(ws \cdot 1000))}$$

The water solubility is part of the $\rightarrow$ Substance Data [Page 221].

10.1.4
Estimations of the Partition Coefficient Octanol/Water

**From water solubility, melting point and molar mass;
by Yalkowski and Valvani (1980)**

The partition coefficient of octanol to water, $\log K_{OW}$, is estimated using the water solubility, ws, the melting point, mp, and the molar mass, M. It is possible because $\log K_{OW}$ is negatively correlated with the water solubility.

$$\log K_{OW} \;=\; 0.95 \cdot \log\left(\frac{ws}{M}\right) + 0.83 - 0.0114 \cdot f(mp - 273.15)$$

$$f(x) \;=\; \begin{cases} x & \text{for } x > 25 \\ 25 & \text{else} \end{cases}$$

Water solubility, melting point and molar mass belong to the →Substance Data [Page 221].

The reverse estimation function is the →Estimation of water solubility by Yalkowski and Valvani [Page 265].

By Karickhoff (1981)

The partition coefficient of octanol to water, $\log K_{OW}$, is estimated using K_{OC}. This equation is the inversion of the →Estimaton of K_{OC} by Karickhoff [Page 262]. A hint will be prompted if K_{OC} is not within [4.11;215 695 8].

$$\log K_{OW} = \log \frac{K_{OC}}{0.411}$$

K_{OC} is part of the →Substance Data [Page 221].

By Schwarzenbach and Westall (1981)

The partition coefficient of octanol to water, $\log K_{OW}$, is estimated using K_{OC}. This equation is the inversion of the →Estimation of K_{OC} by Schwarzenbach and Westall [Page 262].

$$\log K_{OW} = \frac{\log(K_{OC}) - 0.49}{0.72}$$

K_{OC} is part of the →Substance Data [Page 221].

10.1.5
Estimation of BCF

Differing from K_{AW} and K_d, there is no automatic ph-correction for estimated *BCF*s when the compound is dissociating, because the *pH* of the fish may differ from that in the surrounding water.

By Veith

The bioconcentration factor for biota, *BCF*, is estimated using $\log K_{OW}$. This relation (N = 84; $r^2 = 0.82$) was derived by Veith and is applicable mainly for chlorinated hydrocarbons. This estimation function prompts a hint if $\log K_{OW}$ is not within [0.89;6.90].

$$BCF = 10^{0.76 \cdot \log K_{OW} - 0.23}$$

The partition coefficient $\log K_{OW}$ is part of the $\rightarrow$ Substance Data [Page 221].

By Isnard and Lambert (1988)

The bioconcentration factor for biota, *BCF*, is estimated using $\log K_{OW}$. Isnard and Lambert (1988) came to this relation from extensive literature data (N = 107; $r^2 = 0.82$). If the $\log K_{OW}$ is outside [0.98;6.89], this relation should not be applied and the estimation function prompts a warning. The same relation, but based on water solubility is used within the estimation function $\rightarrow$ of BCF from water solubility by Isnard and Lambert [Page 264].

$$BCF = 10^{0.8 \cdot \log K_{OW} - 0.52}$$

The partition coefficient $\log K_{OW}$ is part of the $\rightarrow$ Substance Data [Page 221].

From water solubility; by Isnard and Lambert (1988)

The bioconcentration factor for biota, *BCF*, is estimated using the water solubility, *ws*. Isnard and Lambert came to this relation by extensive literature data (N = 107; $r^2 = 0.75$). If the water solubility is outside [$2 \cdot 10^{-7}$; 36.308], this relation should not be applied and a warning will be prompted. The same relation, but based on $\log K_{OW}$ is used within the estimation function $\rightarrow$ of BCF by Isnard and Lambert [Page 264].

$$BCF = 10^{3.13 - 0.51 \cdot \log(ws \cdot 1000)}$$

The water solubility is part of the $\rightarrow$ Substance Data [Page 221].

10.1.6
Estimation of Diffusion Coefficient for Air from Molar Mass
and Steam as a Reference

The molecular diffusion coefficient with steam as a reference, D_a, is estimated using the molar mass, M, and data for steam.

$$D_a = 2.22 \cdot \sqrt{\frac{18}{M}}$$

Molar mass is part of the →Substance Data [Page 221].

10.1.7
Estimation of Diffusion Coefficient for Water from Molar Mass with Oxygen as a Reference

The molecular diffusion coefficient with oxygen as a reference, D_w, is estimated using the molar mass, M, and data for oxygen.

$$D_w = 1.728 \cdot 10^{-4} \cdot \sqrt{\frac{32}{M}}$$

Molar mass is part of the →Substance Data [Page 221].

10.1.8
Estimation of Water Solubility by Yalkowski and Valvani (1980)

Water solubility, ws, is estimated using $\log K_{OW}$, melting point, mp, and molar mass, M. This is possible because the water solubility is negatively correlated with $\log K_{OW}$. An extensive analysis showed that this relation fails for some chemical classes with OH-group (mainly phenoles).

$$
\begin{aligned}
ws &= 10^{-1.05 \cdot \log K_{OW} + 0.87 - 0.012 \cdot f(mp - 273.15)} \cdot M \\
f(x) &= \begin{cases} x & \text{if } x > 25 \\ 25 & \text{otherwise} \end{cases}
\end{aligned}
$$

Melting point, $\log K_{OW}$ and molar mass are →Substance Data [Page 221].
    The reverse estimation function is the →estimation of the partition coefficient octanol/water by Yalkowski und Valvani [Page 263].

10.2
Estimations of Model Parameters

For a description of practical and automatic estimation see →Estimate Input Data [Page 210].
    In the more complex estimation functions not all parameters are explained. They can be found quickely following the references to the specific models using this method.

10.2.1
Estimation of K_d from OC and K_{OC}

K_d is the partition coefficient of the soil matrix and the soil water. Karickhoff (1981) states that the sorption of hydrophobic organic chemicals to the soil matrix is proportional to the organic carbon content:

$$K_d = K_{OC} \cdot OC$$

Afterwards the $\rightarrow K_d$ standard correction will automatically be carried out [Page 275]. This estimation can be used in the models $\rightarrow$ Water [Page 257], $\rightarrow$ Plant [Page 245], $\rightarrow$ Soil [Page 254], $\rightarrow$ Level1 [Page 234], $\rightarrow$ Level2 [Page 237] and $\rightarrow$ Buckets [Page 227]. K_{OC} is part of the $\rightarrow$ Substance Data [Page 221].

10.2.2
Estimation of Gaseous Deposition Velocity from Molar Mass

The deposition velocity is assumed as a diffusion process. It is computed using the molar mass, M, and data of a reference substance.

$$V_{d,gas} = V_{d,gas}(ref)\sqrt{\frac{M(ref)}{M}}$$

The used reference substance has the following values (Thompson, 1983):

$$M(ref) \quad = \quad 300\,g/mol$$
$$V_{d,gas}(ref) \quad = \quad 5 \cdot 10^{-3}\,m/s$$

This estimation can be used in the models $\rightarrow$ Air [Page 224] and $\rightarrow$ Plume [Page 250]. The molar mass is part of the $\rightarrow$ Substance Data [Page 221].

10.2.3
Estimation of Volatility Rates

This is the base equation for the volatility of the dissolved substance K_V following the two-film theory by Whitman (1923):

$$\frac{1}{K_V} = \left(\frac{1}{k_l} + \frac{1}{K_{AW} \cdot k_g}\right) \cdot d$$

with

k_l : Transfer velocity through the water film
k_g : Transfer velocity through the air film
d : Depth

The effective volatility rate is computed taking sorption to particles into consideration.

$$volat = \frac{K_V}{1 + susp_{matter} \cdot 10^{-3} \cdot K_d}$$

The volatility estimations will compute k_l and k_g in different ways. K_{AW} is part of the $\rightarrow$Substance Data [Page 221].

For flowing waters; by Southworth

This estimation by Southworth (1979) is based on the method of reaeration by Churchill and on additional data by Liss. But first the $\rightarrow K_{AW}$ standard correction is carried out [Page 275]. The first equation computes the wind velocity at a height of 0.1 m from the velocity at a height of 10 m. This is done by the $\rightarrow$Logarithmic wind profile [Page 277]. The roughness height of the water surface is assumed as 1 mm.

$$v_{wind} = windprof(v_{wind10}; 10; 0.1; 0.001)$$

$$k_g = 11.37(v_{wind} + v_{flow})\sqrt{\frac{18}{M}}$$

$$k_l = 0.2351 \cdot v_{flow}^{0.969} d^{-0.673} \sqrt{\frac{32}{M}} \cdot F$$

$$F = \begin{cases} 1 & \text{if } v_{wind} < 1.9\,\text{m/s} \\ e^{0.526(v_{wind}-1.9)} & \text{else} \end{cases}$$

The molar mass M is part of the $\rightarrow$Substance Data [Page 221]. The transfer velocities k_g and k_l now have to be inserted into the main equation of the $\rightarrow$Volatility Estimations [Page 266].

Errors/Warnings:

If the wind velocity is higher than 5 m/s at a height of 10 cm or if the flow velocity is not within [0.5;2] m/s or the depth is outside [3;5] m, this estimation should not be used. A too-high wind speed will cause termination of this estimation. In other cases warnings will be prompted.

If the K_{AW} of a substance is 0, a warning will be prompted and you are asked wheather volatility should be set to 0 or left unchanged.

This estimation can be used in the model $\rightarrow$Water [Page 257].

For Oceans; by Liss and Slater

This volatility model is based on the assumption that the evaporation of water is controlled by air and the evaporation of carbon dioxide by water. The authors gave average values for open oceans: For water k_g (H_2O) = 30 m/h and for carbon dioxide k_l (CO_2) = 0.2 m/h. First the $\rightarrow K_{AW}$ Standard correction will be executed [Page 275].

For chemicals with different molecular weights k_g and k_l have to be multiplied with the fraction of the roots of the molecular weights:

$$k_g = k_g(H_2O)\sqrt{\frac{18}{M}}$$

$$k_l = k_l(CO_2)\sqrt{\frac{44}{M}}$$

The molar mass M is part of the $\rightarrow$ Substance Data [Page 221]. The transfer velocities now have to be inserted into the main equation of the $\rightarrow$ Volatility Estimations [Page 266].

Errors/Warnings:

If the K_{AW} of the substance is 0, a warning will be prompted and you are asked whether volatility should be set to 0 or left unchanged.

This estimation function can be used in the $\rightarrow$ Water [Page 257] model.

For lakes; by Mackay and Yeun

Mackay and Yeun developed an equation combining the Schmidt number, sheer velocity of wind and the transfer velocities k_g and k_l. First the $\rightarrow K_{AW}$ Standard correction will be executed [Page 275].

$$k_g = 3.6 + 166.32 \cdot u \cdot SC_g^{-0.67}$$

$$k_l = \begin{cases} 3.6 \cdot 10^{-3} + 51.84 \cdot u^{2.2} \cdot SC_w^{-0.5} & \text{if } u < 0.3 \\ 3.6 \cdot 10^{-3} + 12.27 \cdot u \cdot SC_w^{-0.5} & \text{if } 0.3 \leq u \leq 1 \end{cases}$$

The Schmidt numbers are defined as.

$$SC_w = \frac{V_w}{D_w}, \quad SC_g = \frac{V_g}{D_g}$$

with

V_w = $1.0 \cdot 10^{-6}\,m^2/s$ (viscosity of water at approx. 18°C; Weast and Astle (1983))

V_g = $14 \cdot 10^{-6}\,m^2/s$ (viscosity of air at 18°C; Weast and Astle (1983))

and the diffusion coefficient for water and air (D_w and D_g) of the $\rightarrow$ Substance Data [Page 221].

The sheer velocity u is calculated by a regression equation of Monin and Yaglom:

$$u = 0.0359 \cdot v_{Wind10}^{0.93}$$

The transfer velocities k_g and k_l now have to be inserted into the main equation of the $\rightarrow$ Volatility Estimations [Page 266].

Errors/Warnings:

If the K_{AW} of a substance is 0, a warning will be prompted and you are asked whether the volatility should set be to 0 or left unchanged.

This estimation can be used in the →Water [Page 257] model.

10.2.4
Estimation of Advective Transport Velocity in Soil

This function estimates the advective transport velocity of a substance in soil. It is used by the →Soil [Page 254] model. Used notations are explained in the description of the model. At the begining the →K_{AW} standard correction will be executed [Page 275].

$$u = (Precipitation - Evaporation - Runoff) \cdot 10^{-3}/pore$$
$$OM = 1.724 \cdot OC$$
$$Density = (1 - Porosity) \cdot (OM \cdot 1400 + (1 - OM) \cdot 2650)$$
$$K_{MW} = K_d \cdot Density \cdot 10^{-3}$$
$$f_W = \frac{pore}{K_{MW} + pore + (Porosity - pore) \cdot K_{AW}}$$
$$u = u \cdot f_W$$

K_{AW} belongs to →Substance Data [Page 221].

10.2.5
Estimation of the Total Dispersion Coefficient in Soil

This function estimates the total diffusion/dispersion coefficient of the substance with the weighting of the fractions of the substance in aquaeous phase and gaseous phase. It is used by →Soil [Page 254] model. Used notations are explained in the description of the model. At the begining the →K_{AW} standard correction will be executed [Page 275].

$$q = (Precipitation - Evaporation - Runoff) \cdot 10^{-3}$$
$$D_{disp} = L_{disp} \cdot q$$
$$D_{W,eff} = D_w \cdot \frac{pore^{\frac{10}{3}}}{Porosity^2}$$
$$D_{G,eff} = D_a \cdot \frac{(Porosity - pore)^{\frac{10}{3}}}{Porosity^2}$$
$$D_{W,a} = D_{disp} + D_{W,eff}$$
$$OM = 1.724 \cdot OC$$
$$Density = (1 - Porosity) \cdot (OM \cdot 1400 + (1 - OM) \cdot 2650)$$

$$K_{MW} = K_d \cdot Density \cdot 10^{-3}$$

$$f_W = \frac{pore}{K_{MW} + pore + (Porosity - pore) \cdot K_{AW}}$$

$$f_G = \frac{(Porosity - pore) \cdot K_{AW}}{K_{MW} + pore + (Porosity - pore) \cdot K_{AW}}$$

$$D = D_{W,a} \cdot f_w + D_{G,eff} \cdot f_g$$

K_{AW}, and D_w and D_a belong to $\rightarrow$Substance Data [Page 221].

10.2.6
Estimations of the Transpiration Stream Concentration Factor

By Briggs (1982)

This function estimates the TSCF (Transpiration Stream Concentration Factor), according to $\rightarrow$Briggs (1982) [Page 270]. The partition coefficient $\log K_{OW}$ is used from the $\rightarrow$Substance Data [Page 221].

$$TSCF = 0.784 \cdot e^{\frac{-(\log K_{OW} - 1.78)^2}{2.44}}$$

This estimation can be used within the $\rightarrow$Plant [Page 245] model.

By Hsu et al. (1990)

This function estimates the TSCF (Transpiration Stream Concentration Factor), according to $\rightarrow$Hsu et al. (1990) [Page 270]. It was found for cinmethylin and related compounds in detopped soybean plants using the pressure chamber technique. The partition coefficient $\log K_{OW}$ is used from the $\rightarrow$Substance Data [Page 221].

$$TSCF = 0.7 \cdot e^{\frac{-(\log K_{OW} - 3.07)^2}{2.78}}$$

This estimation can be used within the $\rightarrow$Plant [Page 245] model.

10.2.7
Estimation of Particle Fraction

For the estimation of the particle fraction (f_p) see $\rightarrow$Junge (1977) [Page 270], who offers the following equation.

$$K_{gp} = \frac{VP}{c \cdot S}$$

$$f_p = \frac{1}{1 + K_{gp}}$$

VP : vapor pressure of substance [Pa]

c : constant 17 [Pa cm]

S : average surface of aerosols ($= 1.5 \cdot 10^{-6}$ [cm^2/cm^3]; i.e. the average background value of the USA)

K_{gp} : partition coefficient gas/particle This estimation can be used in the models $\rightarrow$ Air [Page 224], $\rightarrow$ Plant [Page 245] and $\rightarrow$ Plume [Page 250]. The vapor pressure is part of the $\rightarrow$ Substance Data [Page 221].

10.2.8
Estimation of Concentration in the Soilwater

The concentration in the soilwater (C_{sw}) is estimated using the density of the soil (D_{soil}), the concentration in the soil (C_{soil}), the porosity (Porosity), the fraction of water-filled pores of the soil (pore) and the K_d-value within the model $\rightarrow$ Plant [Page 245]. The density of water (D_{water}) is assumed as 1000 kg/m^3. At the begining the $\rightarrow K_{AW}$ standard correction will be done [Page 275].

$$C_{sw} = \frac{C_{soil}}{K_d \cdot \frac{D_{soil}}{D_{water}} + pore + (Porosity - pore) \cdot K_{AW}}$$

This estimation can be used within the model $\rightarrow$ Plant [Page 245].

K_{AW} is part of the $\rightarrow$ Substance Data [Page 221].

10.2.9
Estimations of the Partition Coefficient Roots/Water

From lipid-, water-content, root density and log K_{OW}

The partition coefficient roots/water (K_{RW}) is estimated using the lipid- and water-content and the density of the roots (L_{roots}, $Water_{roots}$ and D_{roots}). The density of water (D_{water}) is assumed as 1000 kg/m^3 and the density of octanol ($D_{octanol}$) as 822 kg/m^3. a is a correcting factor.

$$K_{RW} = \left(Water_{roots} + L_{roots} \cdot a \cdot (10^{\log K_{OW}})^{0.77} \right) \cdot \frac{D_{roots}}{D_{water}}$$

$$a = \frac{D_{water}}{D_{octanol}} \cdot 1$$

The partition coefficient log K_{OW} is part of the $\rightarrow$ Substance Data [Page 221]. This Estimation can be used within the $\rightarrow$ Plant [Page 245] model.

From root density and log K_{OW}; accordung to Briggs et al. (1982)

The partition coefficient roots/water (K_{RW}) is estimated using the density of the roots (D_{roots}). The density of water (D_{water}) is assumed as 1000 kg/m^3. The equation was given by → Briggs et al. (1982) [Page 167]. It was developed with macerated barley shoots ($n = 7$; $r^2 = 0.96$) and is mainly valid for phenyl carbamides in the range of $-0.7 \leq \log K_{OW} \leq 4.3$. If $\log K_{OW}$ is outside this range, a warning will be given.

$$K_{RW} = \left(0.82 + 0.03 \cdot (10^{\log K_{OW}})^{0.77}\right) \cdot \frac{D_{roots}}{D_{water}}$$

The partition coefficient $\log K_{OW}$ is part of the → Substance Data [Page 221]. This estimation can be used within the → Plant [Page 245] model.

From root density and log K$_{OW}$; by Trapp und Pussemier (1991)

The partition coefficient roots/water (K_{RW}) is estimated using the density of the roots (D_{roots}). The density of water (D_{water}) is assumed as 1000 kg/m^3. Found for Carbamate ($n = 12$; $r^2 = 0.92$) with cut bean-roots und -stems. If $\log K_{OW}$ is outside]1.16; 3.21[the estimation function will generate a warning. The partition coefficient $\log K_{OW}$ is also used (see → Substance Data [Page 221]).

$$K_{RW} = \left(0.85 + 0.046 \cdot (10^{\log K_{OW}})^{0.557}\right) \cdot \frac{D_{roots}}{D_{water}}$$

This estimation can be used within the → Plant [Page 245] model.

10.2.10
Estimation of Partition Coefficient Leaves/Atmosphere

From lipid-, water-content, plant density, log K$_{OW}$ and K$_{AW}$; by Briggs et al. (1983)

The partition coefficient leaves/atmosphere (K_{LA}) is estimated using lipid-, water-content and the density of the surface parts of the plant (L_{plant}, $Water_{plant}$ und D_{plant}) from within the → Plant [Page 245] model. The density of water (D_{water}) is assumed as 1000 kg/m^3 and the density of octanol ($D_{octanol}$) as 822 kg/m^3. a is a correcting factor. The partition coefficient $\log K_{OW}$ and the K_{AW} also have to be known (see → Substance Data [Page 221]).

$$K_{LW} = \left(Water_{plant} + L_{plant} \cdot a \cdot (10^{\log K_{OW}})^{0.95}\right) \cdot \frac{D_{plant}}{D_{water}}$$

$$a = \frac{D_{water}}{D_{octanol}} \cdot 1$$

$$K_{LA} = \frac{K_{LW}}{K_{AW}}$$

K_{LW} represents the partition coefficient leaves/water. The estimation function can only be applied if the K_{AW} is not zero; it will otherwise generate an error.

From density of plant, log K_{OW} and K_{AW}; by Briggs et al. (1983)

The partition coefficient leaves/atmosphere (K_{LA}) is estimated using the surface plant parts from within the model $\rightarrow$ Plant [Page 245]. The density of water (D_{water}) is assumed as 1000 kg/m³. This estimation contains the estimation of SXCF (stem xylem concentration factor) by $\rightarrow$ (Briggs et al. (1983)) [Page 167]. It was developed with macerated barley shoots ($n = 7$; $r^2 = 0.96$) and is mainly valid for phenyl ureas, in the range of $-0.7 \leq \log K_{OW} \leq 4.3$. If $\log K_{OW}$ is outside this range, a warning will be given. K_{AW} and $\log K_{OW}$ also have to be known (see $\rightarrow$ Substance Data [Page 221]).

$$K_{LW} = \left(0.82 + 0.0089 \cdot (10^{\log K_{OW}})^{0.95}\right) \cdot \frac{D_{plant}}{D_{water}}$$

$$K_{LA} = \frac{K_{LW}}{K_{AW}}$$

K_{LW} represents the partition coefficient leaves/water. The estimation function can only be applied if K_{AW} is not zero; it will otherwise generate an error.

10.2.11
Estimation of Partition Coefficient Plant/Water

The partition coefficient plant/water (K_{PW}) is estimated using lipid- $(Plant.L_p)$, water-content $(Plant.W_p)$ and the density of the surface plant parts $(Plant.Density)$. The density of water (D_{water}) is assumed as 1000 kg/m³ and the density of octanol $(D_{octanol})$ as 822 kg/m³.

$$D_w = 1000$$

$$D_{oc} = 822$$

$$a = \frac{D_w}{D_{oc}}$$

$$K_{PW} = \left(Plant.Wp + Plant.Lp \cdot a \cdot (10^{\log K_{OW}})^{0.95}\right) \cdot \frac{Plant.Density}{D_w}$$

The partition coefficient $\log K_{OW}$ belongs to the $\rightarrow$ Substance Data [Page 221]. This estimation can be used within the models $\rightarrow$ Level1 [Page 234] and $\rightarrow$ Level2 [Page 237].

10.2.12
Estimation of Partition Coefficient Sediment/Water

The partition coefficient sediment/water (K_{SW}) is estimated using K_d, the density of the sediment $(Sediment.Density)$ and the water content of the sediment $(Sediment.theta)$.

$$K_{SW} = \frac{Sediment.Density}{1000} \cdot Sediment.K_d + Sediment.theta$$

This estimation can be used within the models →Level1 [Page 234] and →Level2 [Page 237].

10.2.13
Estimation of the Partition Coefficient Soil/Water

The partition coefficient soil/water (K_{BW}) is estimated using the K_d, the density of the soil, the water content of the pores of soil (*Sediment.theta*) and the total fraction of pores (*Sediment.epsilon*).

$$\begin{aligned} K_{BW} \quad = \quad & \frac{Soil.Density}{1000} \cdot Soil.K_d + Soil.theta \\ & +(Soil.epsilon - Soil.theta) \cdot K_{AW} \end{aligned}$$

K_{AW} is part of the →Substance Data [Page 221]. This estimation can be used within the models →Level1 [Page 234] and →Level2 [Page 237].

Chapter 11:
Standard Procedures

In this chapter all standard procedures are explained. These are used directly from estimation functions or models and can be described as small, complete multiply used sub-procedures.

11.1
Standard Correction of K_{AW}

The partition coefficient between air and water, K_{AW}, which is often used in exposure models is dependent on the pH-value if the substance is dissociating. In this case, the K_{AW} is corrected for the part of the neutral molecules. The following correction will always be made if a pH-value occurs in a model. It is the Henderson-Hasselbalch equation.

IF pH exists **AND** K_{AW} is estimated **AND** pKa is known **AND** protolysis of the substance is not neutral, **THEN**

$$a = \begin{cases} 1 & \text{if protolysis} = \text{acidic} \\ -1 & \text{if protolysis} = \text{alkaline} \end{cases}$$

$$K_{AW} = \frac{K_{AW}}{1 + 10^{a(pH - pKa)}}$$

See → Substance Data [Page 221] for more information.

11.2
Standard Correction of K_d

The partition coefficient between soil matrix and soil water, K_d, which is often used in exposure models is dependent on the pH-value if the substance is not dissociating. In this case, the K_{AW} is corrected for the part of the neutral molecules. The following correction will always be made, if a pH-value occurs in a model. It is the Henderson-Hasselbalch equation.

IF *pH* exists **AND** *pKa* is known **AND** protolysis of the substance is not neutral, **THEN**

$$a = \begin{cases} 1 & \text{if protolysis} = \text{acidic} \\ -1 & \text{if protolysis} = \text{alkaline} \end{cases}$$

$$K_d = \frac{K_d}{1 + 10^{a(pH - pKa)}}$$

See →Substance Data [Page 221] for more information.

11.3
Error Functions erf and erfc

An approximation has been implemented by Abramowitz and Stegun (1972) for the error function *erf*:

$$p := 0.47047$$
$$a_1 := 0.3480242$$
$$a_2 := -0.0958798$$
$$a_3 := 0.7478556$$

$$z = |x|$$
$$b = \frac{1}{1 + p \cdot z}$$

$$erf(z) = \begin{cases} 1 & \text{for } z \geq 5 \\ 1 - (a_1 \cdot b + a_2 \cdot b^2 + a_3 \cdot b^3) \cdot e^{-z^2} & \text{for } z < 5 \end{cases}$$

$$erf(x) = \begin{cases} -erf(z) & \text{for } x < 0 \\ erf(z) & \text{else} \end{cases}$$

The complementary error function *erfc* is computed directly from *erf*:

$$erfc(x) = 1 - erf(x)$$

These error function are used in the model →Soil [Page 254].

11.4
Exponential Function and Logarithm

Since there are limitations in the exactness of computations set by the computer, we describe here the functions e^x, x^y and $\log x$ according to their implementation. This should increase the transparency of comparisons with other computations regarding exponential functions as error sources, since the results can be checked. The procedures are implemented as follows ($exp(x)$ and $\ln(x)$ are the internal procedures of Borland Pascal 7.0):

$$e^x = \begin{cases} exp(x) & \text{for } -11356 \le x \le 11356 \\ 0 & \text{else} \end{cases}$$

$$x^y = \begin{cases} 1 & \text{for } y = 0 \\ x^y & \text{for } x < 0 \text{ and } y > 0 \text{ and } y \text{ integer} \\ \frac{1}{x^{|y|}} & \text{for } x < 0 \text{ and } y < 0 \text{ and } y \text{ integer} \\ unde\!fined & \text{for } x < 0 \text{ and } y \text{ not integer} \\ exp(y \cdot \ln(x)) & \text{for } x > 0 \end{cases}$$

$$\log x = \begin{cases} \frac{\ln x}{\ln 10} & \text{for } x > 0 \\ unde\!fined & \text{else} \end{cases}$$

11.5
Logarithmic Wind Profile

Usually the wind velocity is measured at the height of 10 m. For processes that need the wind velocity at another height, this could be computed using the logarithmic wind profile:

$$windprof(x, h_x, h, z) = x \cdot \frac{\log(h/z)}{\log(h_x/z)}$$

with

x :   given wind velocity
h_x :   height of the given wind velocity
h :   height of the needed wind velocity
z :   roughness height of the air at the boundary layer, default $= 0.001$ m

This function is used in the $\rightarrow$ estimation for flowing waters [Page 267].

Chapter 12:
Technical Remarks

12.1
Program Development

Before starting to develop the program, an object-oriented study was undertaken with the shareware program OOTher. The implementation language was
Borland Pascal 7.0. The use of RCS (Revision Control System) allowed the independent work of five programmers on the source code and guaranteed the
reproducability of all program versions. The TEXinfo package was responsible for the correspondence of the manual and the online-help. Furthermore,
several Unix-tools such sed and awd were used.

12.2
Design Decisions

Before starting the implementation of CemoS there some essential design decisions had to be made.

The desktop interface was implemented with Borlands Turbo Vision for DOS
because of its widespread use. All model equations were verified before they
were implemented. The demand for complete comprehensibility was satisfied
using the following three decisions:

- The manual is indentical wih the online-help and describes all model equations and all physical units used.
- Any data the user enters can be commented upon.
- All hints, warnings and errors which appear are presented in a protocol
  window.

12.3
Hard- and Software Requirements

An IBM-PC or a compatible computer is required, running MS-DOS 3.1 or
higher versions as an operating system. Version 5.0 of MS-DOS is required

for MIF-processing using the MS-DOS batch-files or the demonstration, since MORE, PAUSE and pipes are used. The numeric co-processor will be used, if available. CemoS can be run from the disk drive, but for faster results it should be installed onto a hard disk drive.

All common graphics cards can be used. VGA graphics cards are recommended. At least an EGA/VGA card has to be available, for the high-resolution graphics, see →Menu item Show Graphics [Page 200]. If this is not the case, an appropriate BGI-graphic driver has to copied into the CemoS directory.

CemoS uses code page 850. If this code page is not loaded, some special characters are displayed incorrectly. In order to find out how to activate this codepage, refer to the README.TXT file.

A mouse is recommended.

CemoS.EXE works in the protected mode (for 286 or better CPUs) and takes full advantage of the installed memory (not limited to the 640KByte). CemoSRM.EXE uses the real mode of the CPU, which is the XT compatible mode. Therefore it only can access the conventional memory and has to work with overlays which can be placed in the expanded memory, if available. If you wish to run CemoSRM, remove all unnecessary programmes and drivers (e.g. network driver) from the lower memory so that you can work properly. Information about the contents of the conventional memory can be received by means of the *mem* /c DOS-command.

12.4
Software Remarks

The authors would like to comment on the main advantages and disadvantages of the software which was incorporated to the development of CemoS in order to give information and ideas to all those, who develop programmes for research and teaching.

We used many programmes implemented into the spirit of the GNU project of the FSF (Free Software Foundation). This project secures the free availability of a multitude of good programmes. This is very important for high-quality teaching and research. Our development profited directly from the applied tools.

awk: is a widespread tool on all platforms to modify text files →(Aho; Kernighan; Weinberger; 1988) [Page 167]. We used small awk-scripts as converters and preprocessors in this project. These scripts are generally short and well structured and provide a very effective way of processing asciifiles, in our opinion. We used Michael D. Brennan's 1.1.4 version of MAWK. It is enclosed with CemoS, since MIF-processing is implemented in awk.

gnuplot: is a scientific plot program. It can plot functions and data in several ways. We use gnuplot to visualize the tables in MIF-files and to convert the results into postscript format. We have the version 3.5.1.17 of gnuplot installed for MS-DOS and Unix. The CemoS distribution disk contains an older FAQ-file for gnuplot.

RCS: stands for Revision Control System. If more than one developer is involved in a software project, a tool of this kind should be used. It enables more than one author to change a source, so that there are no multiple, incompatible versions of a file. Furthermore, each state of the software can be reconstructed. The latter is also a good argument for using RCS, even if there is only one programmer. RCS requires some discipline, for which you are rewarded afterwards. We used RCS version 4.2 by Walter Tichy.

TEX, texinfo, makeinfo: is a text-setting and layout system with which the user manual and online help of CemoS were created. TEX was also used for the project protocols which was advantageous in comparison to the usually used "screen-orientated" word processing programs. The typographic quality is usually higher. TEX works like a programming language: you write your document as an ascii-file using formatting commands in your text. Whole macro packages can be written for different applications. One already developed macro package is texinfo, which created our user manual and a gnuinfo file from just one source file. For the CemoS online help we used an awk-script to convert the file to the correct format for the Borland Help Compiler. The latter and makeinfo had to be modified slightly and texinfo was adjusted to fulfil our requirements. We used EmTEX version 3.1415 [3c-beta9], GNU makeinfo version 1.55 and texinfo version 2.108.

vim: is a vi-clone which has some additional useful features compared to its ancestor. This program is a pure text editor. It may be difficult use at first, but the powerful commands enable you after a short introductory period to be more effective than with the "usual" editors. We used the 3.0 version of VIM by Bram Moolenaar.

Gnuplot, emTEX, Makeinfo, Mawk, RCS, texinfo, and VIM are (mainly) freely available software.

12.5
History

Version 1.04:

• The previous versions of CemoS where only released in German.

Version 1.05:

• Some strings for the MIF processing were improved. It is "Concentrations" for Level2, which is more correct as "Concentration balance".
• The included substance database STDSUB.DAB has been improved.
• An TSCF estimation function by Hsu et al. (1990) has been included.

Version 1.06:

- The distribution contains an improved DEMO.TVR file. Thus the demonstration is working again.
- README.TXT now includes another annotation about unpacking the archive.

Version 1.07:

- Language improvements for all english texts. Including the manual and small changes to the standardnames strings and the MIF processing.

Version 1.10:

- Language improvements for all English texts, including the manual and slight changes to the file "stdnames.sn" and the MIF processing.
- Loading of scenario files was improved. In previous versions the estimations, in which you could choose between more than one estimation function, were not imported correctly and resulted in program aborts.
- While loading Plant scenarios, it is now verified that the version used for saving the scenarios contained the model parameter "growth_rate". If you are loading a scenario file created with CemoS versions before 1.10, you will receive a warning that CemoS might fail. Files created with CemoS 1.04 or later can be imported without any problems.
- Importing some parameters of a MIF file without importing the substance, or importing the substance part without the model parameters is now more comfortable. In the cass of a sole substance part, the estimation functions are checked after importing all values. Unnecessary warnings are avoided. If parameter parts are imported, all estimation functions are now recalculated. Until this version only the estimation functions which are dependent on the model parameters were reconcidered. But the imported values which are dependent on the substance were forgotten. Recalculation was not undertaken until the model parameter dialog was opened. With the present CemoS version 1.10 a warning is printed, but the text could be misleading. This is far better than using a wrong value, which is now impossible.
- We were informed that the runtime libraries of Borland Pascal 7.0x can cause a runtime error when starting CemoS on fast PCs. Though we could not reproduce that error, we installed a path to fix the problem. Therefore CemoS should be ready to run stably on faster machines to be developed in the future.

12.6
Known Bugs

Units: At present CemoS when importing does not check the units of particular values, which are given in the MIF-files. It is assumed that the values correspond to those in CemoS.

MIF export/import: In the German version of CemoS, the information about the correlation between dialog comments and dialogs is only exported in German and cannot be reconstructed if you import such a file to the English version. This also applies to the model hint section. When using the German version of CemoS, even if the comments are entered in English, no completely perfect MIF-file can be exported for English. The comments may appear in the first dialog and the results of the model hint section can be reconstructed using CemoS, so it is still comprehensible.

MIF import: You might encounter difficulties when importing MIF-files generated with older german CemoS versions. Some standardnames were wrong and importing will leave the corresponding values "unknown". Use URE-PORT.BAT to process such MIF-files. Usually you can see the missing values that way or you can have a look in the MIF-file itself.

MIF processing for Level2: Concentrations and mobility are exported as balances for Level2. So the MIF processing creates a balance with percentages and a sum for this values. This can be misunderstood as the dialogs for Level2 only show percentages regarding the masses on the screen.

Installation of Codepage 850: Depending on the use of different operating system installations, it might be difficult to install the codepage 850, which CemoS uses to display some special characters. But CemoS should remain functional with common codepages.

Import model parameters without a substance: The estimation functions for substance parameters are also recalculated. When you import model parameters this may lead to misleading warnings. If a model parameter depends on the substance, a wrong value might be imported. But the estimation is redone and a warning is printed. In this warning the "old value" refers to the freshly imported value and not to the value before importing. Therefore it is correct to choose the new estimated value.

We are grateful for any comments about errors and bugs, suggestions for improvements, etc. Every comment will be dealt with thoroughly. Our e-mail address is:

cemos@usf.Uni-Osnabrueck.DE

If you specify "please send automatic information" as the "Subject" of your mail, you will receive the latest information about CemoS.

Index

Springer
and the
environment

At Springer we firmly believe that an
international science publisher has a
special obligation to the environment,
and our corporate policies consistently
reflect this conviction.
We also expect our business partners –
paper mills, printers, packaging
manufacturers, etc. – to commit
themselves to using materials and
production processes that do not harm
the environment. The paper in this
book is made from low- or no-chlorine
pulp and is acid free, in conformance
with international standards for paper
permanency.